Gene therapy

Published by Woodhead Publishing Limited, 2013

Woodhead Publishing Series in Biomedicine

1 Practical leadership for biopharmaceutical executives
J. Y. Chin
2 Outsourcing biopharma R&D to India
P. R. Chowdhury
3 Matlab® in bioscience and biotechnology
L. Burstein
4 Allergens and respiratory pollutants
Edited by M. A. Williams
5 Concepts and techniques in genomics and proteomics
N. Saraswathy and P. Ramalingam
6 An introduction to pharmaceutical sciences
J. Roy
7 Patently innovative: How pharmaceutical firms use emerging patent law to extend monopolies on blockbuster drugs
R. A. Bouchard
8 Therapeutic protein drug products: Practical approaches to formulation in the laboratory, manufacturing and the clinic
Edited by B. K. Meyer
9 A biotech manager's handbook: A practical guide
Edited by M. O'Neill and M. H. Hopkins
10 Clinical research in Asia: Opportunities and challenges
U. Sahoo
11 Therapeutic antibody engineering: Current and future advances driving the strongest growth area in the pharmaceutical industry
W. R. Strohl and L. M. Strohl
12 Commercialising the stem cell sciences
O. Harvey
13 Biobanks: Patents or open science?
A. De Robbio

Published by Woodhead Publishing Limited, 2013

14 Human papillomavirus infections: From the laboratory to clinical practice
F. Cobo

15 Annotating new genes: From *in silico* screening to experimental validation
S. Uchida

16 Open-source software in life science research: Practical solutions in the pharmaceutical industry and beyond
Edited by L. Harland and M. Forster

17 Nanoparticulate drug delivery: A perspective on the transition from laboratory to market
V. Patravale, P. Dandekar and R. Jain

18 Bacterial cellular metabolic systems: Metabolic regulation of a cell system with ^{13}C-metabolic flux analysis
K. Shimizu

19 Contract research and manufacturing services (CRAMS) in India: The business, legal, regulatory and tax environment
M. Antani and G. Gokhale

20 Bioinformatics for biomedical science and clinical applications
K-H. Liang

21 Deterministic versus stochastic modelling in biochemistry and systems biology
P. Lecca, I. Laurenzi and F. Jordan

22 Protein folding *in silico*: Protein folding versus protein structure prediction
I. Roterman

23 Computer-aided vaccine design
T. J. Chuan and S. Ranganathan

24 An introduction to biotechnology
W. T. Godbey

25 RNA interference: Therapeutic developments
T. Novobrantseva, P. Ge and G. Hinkle

26 Patent litigation in the pharmaceutical and biotechnology industries
G. Morgan

27 Clinical research in paediatric psychopharmacology: A practical guide
P. Auby

28 The application of SPC in the pharmaceutical and biotechnology industries
T. Cochrane

Published by Woodhead Publishing Limited, 2013

29 Ultrafiltration for bioprocessing
H. Lutz

30 Therapeutic risk management of medicines
A. K. Banerjee and S. Mayall

31 21st century quality management and good management practices: Value added compliance for the pharmaceutical and biotechnology industry
S. Williams

32 Sterility, sterilisation and sterility assurance for pharmaceuticals
T. Sandle

33 CAPA in the pharmaceutical and biotech industries: How to implement an effective nine step programme
J. Rodriguez

34 Process validation for the production of biopharmaceuticals: Principles and best practice.
A. R. Newcombe and P. Thillaivinayagalingam

35 Clinical trial management: An overview
U. Sahoo and D. Sawant

36 Impact of regulation on drug development
H. Guenter Hennings

37 Lean biomanufacturing
N. J. Smart

38 Marine enzymes for biocatalysis
Edited by A. Trincone

39 Ocular transporters and receptors in the eye: Their role in drug delivery
A. K. Mitra

40 Stem cell bioprocessing: For cellular therapy, diagnostics and drug development
T. G. Fernandes, M. M. Diogo and J. M. S. Cabral

41 Oral delivery of insulin
T. A. Sonia and Chandra P. Sharma

42 Fed-batch fermentation: A practical guide to scalable recombinant protein production in *Escherichia coli*
G. G. Moulton and T. Vedvick

43 The funding of biopharmaceutical research and development
D. R. Williams

44 Formulation tools for pharmaceutical development
Edited by J. E. Aguilar

45 Drug-biomembrane interaction studies: The application of calorimetric techniques
Edited by R. Pignatello

46 Orphan drugs: Understanding the rare drugs market
E. Hernberg-Ståhl
47 Nanoparticle-based approaches to targeting drugs for severe diseases
J. L. Arias
48 Successful biopharmaceutical operations: Driving change
C. Driscoll
49 Electroporation-based therapies for cancer: From basics to clinical applications
Edited by R. Sundararajan
50 Transporters in drug discovery and development: Detailed concepts and best practice
Y. Lai
51 The life-cycle of pharmaceuticals in the environment
R. Braund and B. Peake
52 Computer-aided applications in pharmaceutical technology
Edited by J. Djuris
53 From plant genomics to plant biotechnology
Edited by P. Poltronieri, N. Burbulis and C. Fogher
54 Bioprocess engineering: An introductory engineering and life science approach
K. G. Clarke
55 Quality assurance problem solving and training strategies for success in the pharmaceutical and life science industries
G. Welty
56 Advancement in carrier based drug delivery
S. K. Jain and A. Jain
57 Gene therapy: Potential applications of nanotechnology
S. Nimesh
58 Controlled drug delivery: The role of self-assembling multi-task excipients
M. Mateescu
59 *In silico* protein design
C. M. Frenz
60 Bioinformatics for computer science: Foundations in modern biology
K. Revett
61 Gene expression analysis in the RNA world
J. Q. Clement
62 Computational methods for finding inferential bases in molecular genetics
Q-N. Tran

The Woodhead/Chandos team responsible for publishing this book:
Publisher: Dr Glyn Jones
Production Editor: Ed Gibbons
Project Manager: Annette Wiseman
Copy Editor: Sue Clements
Cover Designer: Ian Hutchins

Woodhead Publishing Series in Biomedicine: Number 57

Gene therapy

Potential applications of nanotechnology

SURENDRA NIMESH

Oxford Cambridge Philadelphia New Delhi

Published by Woodhead Publishing Limited, 2013

Woodhead Publishing Limited, 80 High Street, Sawston, Cambridge, CB22 3HJ, UK
www.woodheadpublishing.com
www.woodheadpublishingonline.com

Woodhead Publishing, 1518 Walnut Street, Suite 1100, Philadelphia, PA 19102-3406, USA

Woodhead Publishing India Private Limited, G-2, Vardaan House, 7/28 Ansari Road, Daryaganj, New Delhi – 110002, India
www.woodheadpublishingindia.com

First published in 2013 by Woodhead Publishing Limited
ISBN: 978–1–907568–40–4 (print); ISBN 978–1–908818–64–5 (online)
Woodhead Publishing Series in Biomedicine ISSN 2050-0289 (print); ISSN 2050-0297 (online)

British Library Cataloguing-in-Publication Data: A catalogue record for this book is available from the British Library.

Library of Congress Control Number: 2013940432

Typeset by RefineCatch Limited, Bungay, Suffolk
Printed in the UK and USA

Dedicated to my beloved daughters,
Sunidhi, Sanvi
and my wife Nidhi

Published by Woodhead Publishing Limited, 2013

Contents

Published by Woodhead Publishing Limited, 2013

Published by Woodhead Publishing Limited, 2013

Published by Woodhead Publishing Limited, 2013

Published by Woodhead Publishing Limited, 2013

List of figures and tables

Figures

Published by Woodhead Publishing Limited, 2013

Published by Woodhead Publishing Limited, 2013

Published by Woodhead Publishing Limited, 2013

Tables

Published by Woodhead Publishing Limited, 2013

Acknowledgments

A work of this scope would be impossible without the dedication and hard work of many people. I sincerely thank the scientists who have devoted their careers to understanding the nanotechnology world, and provide careful explanation and clarity enough for a beginning audience. I wish to express my gratitude to those who generously helped to color the mosaic of this book with the tiles of their knowledge and expertise.

The first and foremost thanks are due to Prof. Ramesh Chandra, Founder Director, Dr. B.R. Ambedkar Center for Biomedical Research, University of Delhi, Delhi, for his persistent interests and generous availability of all his expertise in the field of nanotechnology throughout the course of this work. His optimism and dynamic attitude have been a constant source of inspiration.

The next best but equally heartfelt gratitude goes to Prof. Satya Prakash, Biomedical Technology and Cell Therapy Research Laboratory, McGill University, Montreal, Canada, for his encouragement, motivation, undying patience and critical evaluation are too tremendous to fit into words. He has been a guiding star all along, supporting with his inexhaustible knowledge and immense experience. His invaluable inputs and unfailing encouragement throughout this work is an asset.

Carefully reviewing a book means lots of work but not much appreciation for the reviewers. Therefore, the author

Published by Woodhead Publishing Limited, 2013

wishes to express his deep gratitude to Dr. Nidhi Gupta for taking on this difficult job.

Furthermore, I would like to thank Dr Glyn Jones and the Chandos publishing house for supporting the publication of this book.

The author shall never be able to express, to any degree of satisfaction, his deepest feelings of gratefulness to his friends and whole family for their untiring efforts and for helping to finalize this work.

Dr. Surendra Nimesh

Published by Woodhead Publishing Limited, 2013

Foreword

Globally dubbed as "Technology of the Future", the field of nanoscience and nanotechnology explore distinct properties of materials at nano-scale and exploit them to increase the efficiency of the existing products or create novel ones. Though spanning a range of science and engineering disciplines, the fastest growing technology in the world has particularly promised to revolutionize the area of human health care; the way we diagnose and treat diseases. Gene therapy: potential applications of nanotechnology by Dr. Surendra Nimesh is a comprehensive articulation of basic theory, applications and challenges involved in the use of this inexorable technology in the field of medicine.

Gene therapy also referred to as "gene transfer research" is, in simple words, a technique by which an abnormal or non-functional gene is replaced in the genome by a normal one. At present more than 1800 gene therapy clinical trials are underway world wide (for details of these clinical trials please refer to Ginn et al., J Gene Med 2013; 15: 65-77). Despite the early and exciting progress made towards the utility of gene therapy to treat several diseases including genetic disorders and cancers, one of the perils of the technique has been high rates of death due to the viral vectors used as vehicles for transferring genes. Nanotechnology and nanomaterials offer safe surrogates to the traditional virus-based delivery vehicles and offer tremendous potential in the areas of nanomedicine.

Published by Woodhead Publishing Limited, 2013

This forward-thinking book written by Dr. Surendra Nimesh, is for both nano-specialists and non-nano audience, and provides a detailed technical preface to nanomaterials and their applications, concepts of gene therapy, and critically discusses the advantages and disadvantages associated with the use of nano-vehicles for gene therapy.

Sabina Halappanavar
Research Scientist, Genomics and Nanotoxicology Laboratory
Mechanistic Studies Division
Environmental Health Science and Research Bureau, Health Canada
Tunney's Pasture, P/L 0803A
50 Columbine Driveway, Ottawa, Ontario K1A0K9

Published by Woodhead Publishing Limited, 2013

Preface

Gene therapy is rapidly emerging as a new class of therapeutics for the treatment of inherited and acquired diseases. However, poor cellular uptake and instability of DNA in physiological milieu limits its therapeutic potential, hence there is a need to develop a vector which can protect and efficiently transport DNA to the target cells. Nanotechnology-based non-viral vectors have been proposed as a potential alternate. Various polymeric nanoparticles have been shown as suitable delivery candidates with high cellular uptake efficiencies and the advantage of reduced cytotoxicity. These delivery vectors form condensed complexes with DNA which results in shielding against enzymatic degradation and enhanced cellular targeting. Nanoparticles derived from various polymers like poly-L-lysine, polyethylenimine, and chitosan have been largely explored, as they bear several advantages such as: easy manipulatibility, high stability, low cost and high payload. This new book reviews the research work done in gene delivery employing nanotechnology. Several cationic polymers used to prepare nanoparticles for gene delivery have been discussed with relevant recently published data.

Chapters 1–4 – Provide a brief introduction to nanotechnology and conceptualization of targeted gene and drug delivery. The nanoparticles derived from various sources such as from cationic polymers have been investigated for gene delivery studies. In these chapters, an emphasis has been given on the available tools and techniques for *in vitro*

Published by Woodhead Publishing Limited, 2013

and *in vivo* characterization of nanoparticles. Right from the preparation of nanoparticles till the final fate i.e. release of payload followed by transfection efficiency has been mentioned here. Nanoparticle by virtue of their small size possess excellent gene delivery efficacies however, it also renders difficulty in characterization and hence need of highly sophisticated analytical instrumentation. A wide variety of highly sophisticated techniques such as TEM, SEM, AFM, and DLS has been employed to study nanoparticles.

Chapters 5–7 – Discuss the concept and theory behind nanoparticles mediated gene delivery. Gene delivery is a multi-step process and faces numerous challenges at each and every step. These chapters detail the barriers encountered during gene delivery and the strategies adapted to encounter those using nanoparticles. The cationic nanocarriers have been suggested to not only overcome the cellular and environmental hurdles but also result in targeted and sustained gene delivery. Polymeric nanoparticles decorated with various targeting ligands including antibodies, RGD peptides, cholesterol etc. showed promising results in *in vitro* and *in vivo* studies done in animal models.

Chapters 8–16 – Provide a detailed account of the various nanomaterials employed to deliver DNA for *in vitro* and *in vivo* gene delivery. The initial studies carried out with cationic polymers such as chitosan, PEI, PLL are exciting and need to be further validated before entering clinical application. An in-depth archive has been drawn in these chapters for investigations going on into various polymeric nanoparticles for gene delivery.

Published by Woodhead Publishing Limited, 2013

About the author

Dr. Surendra Nimesh is an internationally recognized expert on nanotechnology for biological applications with specialization in drug and gene delivery. He received his M.S. in Biomedical Science from Dr. B.R. Ambedkar Center for Biomedical Science Research (ACBR), University of Delhi, Delhi, India in 2001. He completed his PhD. in Nanotechnology at ACBR and Institute of Genomics and Integrative Biology (CSIR), Delhi, India in 2007. After completing his postdoctoral studies with Prof. M.D. Bushmann at Ecole Polyetchnique of Montreal, Montreal in 2009, he joined Clinical Research Institute of Montreal, Montreal, Canada as Postdoctoral Fellow. He worked for a short duration with Prof. Satya Prakash at McGill University, Montreal and thereafter joined Health Canada, Canada as NSERC visiting fellow in 2012. He joined Central University of Rajasthan, India as Assistant Professor at School of Life

Published by Woodhead Publishing Limited, 2013

Sciences in 2013. He has authored more than 14 research papers, 5 review articles in international peer reviewed journal, 4 book chapters and 2 books. He has given two invited lectures in international conferences in Turkey in 2010 and presented several papers in international conferences in India and US. His biography has been listed in Marquis Who's Who in Science and Engineering 2011–2012 (11th Edition). He has received Council of Scientific and Industrial Research – University Grants Commission's, India Junior Research Fellowship (JRF) and Eligibility for Lectureship – National Eligibility Test (NET), Dec. 2000 and Young Scientist Project from Department of Science and Technology (SERC Fast Track Proposals for Young Scientists Scheme), Government of India, March 2007 and August 2012. Apart from this, during his doctoral degree and research work, he had experience in training Masters and undergraduate interns and made several presentations in an institute and abroad. His research interests include nanoparticles mediated gene, siRNA and drug delivery for therapeutics.

Published by Woodhead Publishing Limited, 2013

1

Nanotechnology: an introduction

DOI: 10.1533/9781908818645.1

Abstract: The term "nanotechnology" was coined by Norio Taniguchi (1974) and defined as: Nanotechnology mainly consists of the processing of separation, consolidation, and deformation of materials by one atom or one molecule. Nanotechnology enables the production and application of physical, chemical, and biological systems at sizes ranging from individual atoms or molecules to submicron dimensions. Nanomedicine is one of the most significant implications of nanotechnology; it is used to collectively mention liposomes, quantum dots, polymeric micelles, polymer-drug conjugates, dendrimers, inorganic nanoparticles, biodegradable nanoparticles, and other materials in nanoscale size with therapeutic relevance. This chapter provides some basic introduction to nanotechnology and various drug and gene delivery systems.

Key words: nanotechnology, nanomedicine, nanoparticles, nanocapsules, nanospheres, liposomes, quantum dots.

1.1 Introduction

In the quest to develop an efficient drug delivery system with desired therapeutic effect, several approaches have been proposed. Upon administration, the fate of drug distribution in the body depends on the physico-chemical properties and molecular structure of the drug. However, the uneven distribution of the drug results in only a small amount of the administered dose reaching the desired site of action. Further, deposition of the drug at non-specific sites may lead to adverse and toxic side effects. Hence, it has been a challenging task to design and develop novel delivery systems with maximal therapeutic effect and minimal toxic effects. To develop better drug targeting strategies it is imperative to fully explore the nature of the target and the mechanism of targeting. A number of drug delivery systems have been developed in order to engineer a system that can efficiently and specifically deliver drugs to target sites. One early approach was the use of prodrugs, which are modified forms of the active drugs that can reach the target site and be cleaved enzymatically or chemically to release the active drug moiety. More recently, several biomolecule-based vectors have been investigated, including polyclonal antibodies, monoclonal antibodies, sugars and lectins, as targeting molecules to which the drug can be chemically coupled.

The primary objective of controlled drug delivery systems is to deliver a drug to a target site, which could be an organ, a tissue or a particular population of cells within a tissue. Regarding development of newer target-specific delivery systems, the following points are noteworthy:

- the system should be capable of distinguishing between target and non-target sites, to diminish the side effects;

Published by Woodhead Publishing Limited, 2013

- the system should not exhibit toxic effects on prolonged *in vivo* administration;
- the vector should be biodegradable, i.e. it should degrade after achieving drug delivery and the degraded components thus formed should not be toxic.

The site-specific delivery of a drug can be significantly influenced by the choice of delivery system. Recent decades have witnessed the evolution of several carrier systems, such as microparticles, microemulsions, liposomes and nanoparticles. These systems were designed to deliver drug at the target sites and diminish any interactions with non-target sites, thereby leading to enhanced therapeutic effect. These novel carrier systems offer explicit advantages, such as:

- maintenance of drug activity;
- maintenance of therapeutic drug concentration at target sites;
- protection of drug from being eliminated by the host defense system.

The host defense mechanism, which consists of the reticuloendothelial system (RES), poses one of the major challenges in the development of polymeric nanoparticulate vectors. When a carrier molecule, or any other foreign moiety, enters the vascular system, it is rapidly conditioned (or coated) by elements in circulation, such as plasma proteins and glycoproteins called "opsonins", in a process known as "opsonization". This opsonization process enables easy identification of the carrier materials by the RES, as in the case of other foreign bodies, pathogens, or dying cells, and cleared from the blood circulation by phagocytosis by the macrophages.[1] The macrophage cells of the liver (Kupffer cells), spleen and lung, and circulating macrophages, all play

Published by Woodhead Publishing Limited, 2013

a significant role in the removal of the opsonized particles. Recognition of the particles by the RES is dictated by size and surface properties. Particulates with large hydrophobic surfaces are efficiently coated with opsonins and are rapidly cleared from circulation. However, particles with hydrophilic surfaces escape recognition by opsonins, leading to prolonged time in circulation.[2]

Several strategies have been proposed for the modification of surface properties of delivery vectors.[3] Coating of polystyrene nanoparticles with positively-charged polylysyl-gelatin led to the reduction in liver uptake and an increase in spleen and lung uptake. In the early 1980s, Illum and Davis observed that particles can be largely protected from scavenging and clearance by the RES by chemical modification, which consists of coating the particles with block copolymers.[2] This approach proved to be highly effective in alteration of the biodistribution of radiolabeled colloidal particles. The polystyrene particles coated with block copolymer poloxamer 338 largely escaped (by up to 50%) the normally predominant liver and spleen uptake, but could not escape removal by RES cells. However, coating the particles with poloxamine 908, which totally prevents RES capture, led to prolonged circulation times. Some of the other strategies proposed to circumvent these problems include coating the carrier molecules with surfactants, gangliosides and polymers like polyethylene glycol (PEG) or polyethylene oxide (PEO). Studies with surfactant coated carrier molecules led to reduced liver uptake and increase in their concentration in blood and other non-RES organs.[4] However, such strategies are not considered to be applicable in clinical practice, since repeated suppression can lead to impaired RES function.

Particle based drug carriers have been engineered to enhance the bioavailability of drug at the desired site

Published by Woodhead Publishing Limited, 2013

and to explore the therapeutic relevance of the controlled release of drug and drug targeting. It is highly desirable to control the size of the polymeric matrix, to achieve the desired therapeutic response of the entrapped drug molecules. According to size, the particles can be broadly categorized as:

- macroparticles (50–200 μm),
- microparticles (1–50 μm),
- nanoparticles (1–1000 nm).

Microparticles and nanoparticles have been employed in numerous studies of drug and gene delivery. Although microparticles have been observed to circulate in the blood and are capable of passing through the heart, due to their large size they cannot enter capillaries and thus cannot reach tissue sinusoids. However, the drug encapsulated molecules accumulate in the nearby tissues surrounding the capillaries and release the drug slowly. These accumulated microparticles can be used as a depot system and can be delivered by various routes, such as intra-arterial, subcutaneous, intravenous or intraperitoneal. Thus, the drug is not only released slowly and continuously but is also protected from *in vivo* degradation. Efficient drug release is achieved by the use of a polymer properly engineered with regard to its pore size, swelling properties and degradation. Considering these key parameters, biodegradable microparticles of starch, albumin, polylactic acid and ethyl cellulose have been designed and implicated for chemoembolization.

Advances in drug delivery studies have suggested that micron sized particles are rapidly cleared by the RES defense mechanism. However, to circumvent this clearance of particles from the circulation, reduction in size of the particles was thought to help prolong circulation and thereby result in a better carrier for enzymes, proteins or polynucleotides by any route of administration. Further, the decrease in size would result in an increase in surface area, which would allow adsorption of larger amounts of the drug or biological molecule. Subsequently, it has been demonstrated that nanoparticles have the capability of not only releasing a drug or a bioactive molecule at the target site but also carrying it there. The size of the particles had a significant effect, rendering them suitable for systematic application, as they can easily pass through capillaries. Nanoparticles can be defined as metallic or polymeric particles within the size range of 1–1000 nm. Birrenbach and Speiser in 1976 for the first time reported the preparation and characterization of polymeric nanoparticles.[5] This report triggered extensive research into the design and development of novel nanoparticle based carrier systems for delivery of biomolecules. The release of the drug/biomolecule entrapped in the polymeric nanoparticles can be achieved by any one or a combination of the following mechanisms:

- swelling of the polymeric nanoparticle matrix by hydration followed by bursting or slow release through diffusion;
- enzymatic degradation of the polymeric network at the target site, thereby resulting in release of the drug;
- chemical cleavage of the drug molecules from the swollen nanoparticles.

Genes have become potential targets for therapeutic use in a wide variety of diseases due to their ability to produce active proteins using the biosynthetic machinery of the host cell. Gene therapy can be defined as insertion of genetic material, i.e. DNA or RNA, into the cell's genetic machinery either to

Published by Woodhead Publishing Limited, 2013

correct an underlying defect or to modify the characteristics of the cell via expression of the newly inserted gene. However, upon delivery genetic material is rapidly degraded by the nucleases present in the host system. Hence, for efficient delivery of genes to the target cells, several carrier systems have been developed, broadly categorized as viral and non-viral vectors. The viral vectors were initially proposed as potential carriers due to site specificity and the natural mechanism employed to transit cargo to the cells, but their wide clinical application is hampered by immunogenicity and pathogenicity.[6] More recently, a wide variety of non-viral vectors, consisting of lipid based carriers, polycationic lipids, poly-L-lysine (PLL), polyornithine, histones and other chromosomal proteins, hydrogel polymers, calcium phosphate nanoparticles and cationic polymers have been developed. Although non-viral vectors have several advantages, such as ease of manipulation, biodegradability, biocompatibility, etc., they also suffer from some limitations. More specifically, polymeric nanoparticles possess high stability in the endosomal compartment, which can lead to the destruction of DNA by the lysosomal enzymes. Further, high stability of the complexes not only makes DNA more prone to enzymatic degradation, but also inhibits the release of DNA from the complexes within the endosome into the cytosol, thereby resulting in poor transfection efficiency.

1.2 Definition of nanotechnology

A presentation at the annual meeting of the American Physical Society on 29 December 1959, at the California Institute of Technology, by Nobel Laureate Richard P. Feynman, titled "There's Plenty of Room at the Bottom,"

Published by Woodhead Publishing Limited, 2013

has become one of the twentieth century's classic science lectures. Feynman envisaged a technology of manipulating and controlling things on a miniature scale, constructing nano-objects atom by atom or molecule by molecule. The term "nanotechnology" was coined in 1974 by Norio Taniguchi, a professor at Tokyo Science University, and defined as follows: "Nano-technology mainly consists of the processing of separation, consolidation, and deformation of materials by one atom or one molecule." Furthermore, nanotechnology refers to technology operated on a nanoscale with potential applications (Figure 1.1). Nanotechnology enables the production and application of physical, chemical, and biological systems at sizes ranging from individual atoms or molecules to submicron dimensions. Innovation in nanotechnology promises breakthroughs in areas such as materials and manufacturing, nanoelectronics, medicine and healthcare, energy, biotechnology, information technology, and national security. One of the unique aspects of

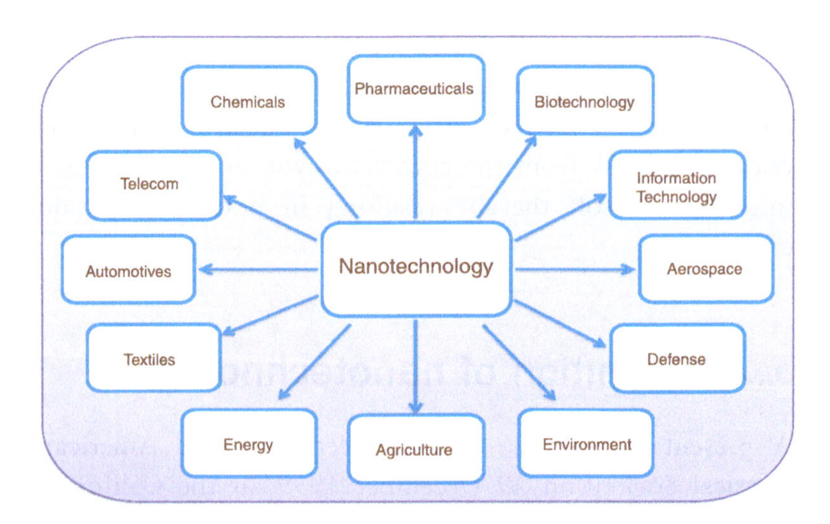

Figure 1.1 Potential applications of nanotechnology

Published by Woodhead Publishing Limited, 2013

nanotechnology is the large increase in the ratio of surface area to volume present in many nanoscale materials, which opens new avenues in surface-based science, such as catalysis.

Nanomedicine is one of the most significant implications of nanotechnology; it is used to collectively describe liposomes, quantum dots, polymeric micelles, polymer–drug conjugates, dendrimers, inorganic nanoparticles, biodegradable nanoparticles, and other materials of nanoscale size with therapeutic relevance. Nanoparticles have emerged as one of the most potential candidates for nanomedicine, with several applications in targeted drug and gene delivery. Nanoparticles provide better tissue penetration and targeting due to their small size.[7] Nanoparticles prepared from polycationic polymers have been largely explored to deliver DNA and small interfering RNA (siRNA). These nanoparticles can be subdivided into (i) nanospheres, spherical nanometer size particles in which the desired molecules can be either entrapped inside the sphere or adsorbed on the outer surface or both; (ii) nanocapsules, which have a solid polymeric shell and an inner liquid core, and the desired molecules can be either entrapped inside the core or adsorbed on the outer surface or both (Figure 1.2). Further, nanoparticles have also been reported to exist in different types of shapes, such as nanorods, nanotubes, cones, spheroids, etc. Recently, carbon nanotubes have also been conjugated with polyglycerol to prepare nanocapsules.[8] Several natural and synthetic polycationic polymers, including chitosan and polyethylenimine (PEI), have been employed to deliver DNA in the form of either complexes or nanoparticles. Herein, I have focussed on recent strategies and advances in the application of polymeric nanoparticles for targeted DNA delivery.

Published by Woodhead Publishing Limited, 2013

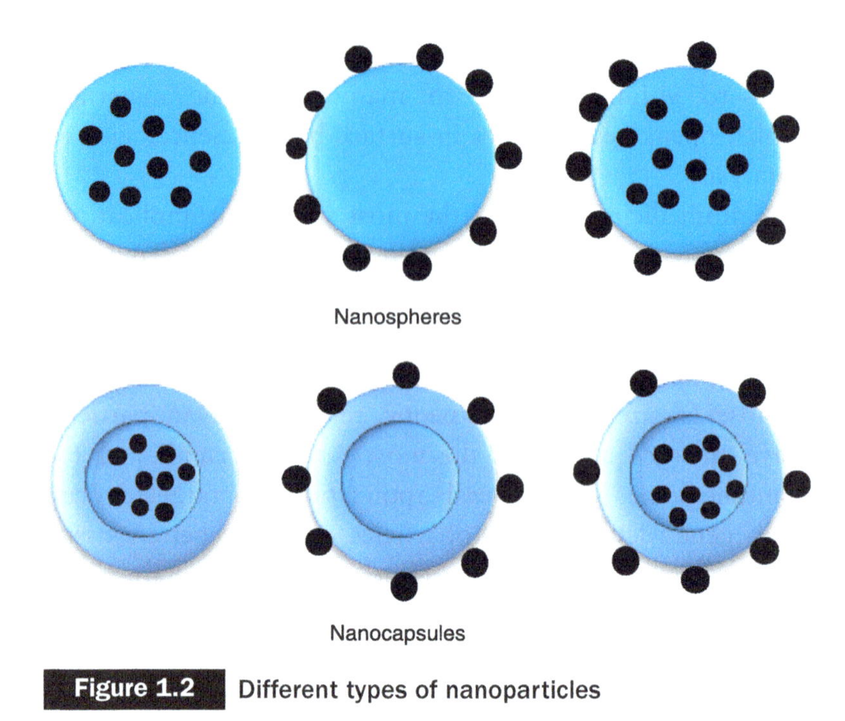

Figure 1.2 Different types of nanoparticles

1.3 Structure of the book

The book compiles basic knowledge of key concepts in preparation, characterization and biological applications of nanoparticles. The first four chapters provide a detailed account of knowledge about the various tools and techniques available for characterization of nanoparticles. The following chapters talk about the basic science involved behind gene therapy, the major barriers encountered during *in vitro* and *in vivo* gene delivery, and the strategies adapted to circumvent them. The second half of the book provides an overview of the various polymers, such as PLL, chitosan, PEI, CDs (cyclodextrins), etc., employed to deliver DNA for gene therapy. The emphasis is on the

Published by Woodhead Publishing Limited, 2013

incorporation of the most up-to-date available published data for both *in vitro* and *in vivo* gene therapy mediated by various polymeric nanoparticles.

1.4 References

1. Müller, R.H. and Wallis, K.H. (1993) Surface modification of i.v. injectable biodegradable nanoparticles with poloxamer polymers and poloxamine 908. *Int J Pharm*, **89**, 25–31.
2. Illum, L. and Davis, S.S. (1984) The organ uptake of intravenously administered colloidal particles can be altered using a non-ionic surfactant (Poloxamer 338). *FEBS Lett*, **167**, 79–82.
3. Wilkins, D.J. and Myers, P.A. (1966) Studies on the relationship between the electrophoretic properties of colloids and their blood clearance and organ distribution in the rat. *Br J Exp Pathol*, **47**, 568–76.
4. Vittaz, M., Bazile, D., Spenlehauer, G., Verrecchia, T., Veillard, M., et al. (1996) Effect of PEO surface density on long-circulating PLA-PEO nanoparticles which are very low complement activators. *Biomaterials*, **17**, 1575–81.
5. Birrenbach, G. and Speiser, P.P. (1976) Polymerized micelles and their use as adjuvants in immunology. *J Pharm Sci*, **65**, 1763–6.
6. Yang, Y., Nunes, F.A., Berencsi, K., Furth, E.E., Gonczol, E. et al. (1994) Cellular immunity to viral antigens limits E1-deleted adenoviruses for gene therapy. *Proc Natl Acad Sci U S A*, **91**, 4407–11.
7. Peer, D., Karp, J.M., Hong, S., Farokhzad, O.C., Margalit, R. et al. (2007) Nanocarriers as an

Published by Woodhead Publishing Limited, 2013

emerging platform for cancer therapy. *Nat Nano*, 2, 751–60.

8. Adeli, M., Mirab, N. and Zabihi, F. (2009) Nanocapsules based on carbon nanotubes-graft-polyglycerol hybrid materials. *Nanotechnology*, 20, 485603.

Published by Woodhead Publishing Limited, 2013

2

Methods of nanoparticle preparation

DOI: 10.1533/9781908818645.13

Abstract: Nanoparticles have been largely prepared from polymers, as they bear several advantages such as high reproducibility, control over physico-chemical properties, low cost, high stability, and wide availability of biodegradable polymers. Polymeric nanoparticles can be fabricated in two different ways, i.e. by polymerization of monomers or from preformed polymers. Various methodologies have been proposed for the preparation of nanoparticles depending upon the type of molecules to be delivered. Polymeric nanoparticles can be prepared by polymerization of monomers using various polymerization techniques such as emulsion, dispersion, and interfacial polymerization. Further, nanoparticles from polymers can be prepared by different approaches including solvent evaporation, salting-out, thermal denaturation, dialysis and supercritical fluid technology. This chapter outlines the different strategies for the preparation of polymeric nanoparticles employed for drug and gene delivery.

Key words: polymeric nanoparticles, monomers, emulsion polymerization, dispersion polymerization, interfacial polymerization, solvent evaporation, salting-out, dialysis.

Published by Woodhead Publishing Limited, 2013

2.1 Introduction

Polymers are attractive candidates for the preparation of nanoparticles as they bear several advantages, such as high reproducibility, control over physico-chemical properties, low cost, high stability, and wide availability of biodegradable polymers. In order to achieve nanoparticles with tailored physico-chemical properties, the method of preparation plays a pivotal role. Nanoparticles of pharmaceutical relevance were first reported by Birrenbach and Speiser in 1976.[1] Polyacrylamide nanoparticles were synthesized, by polymerization of acrylamide cross-linked with N,N´-methylenebisacrylamide (MBA) in an inverse microemulsion (water-in-oil) reaction. Kreuter and Speiser proposed another strategy for the preparation of poly(methyl methacrylate) (PMMA) nanoparticles called dispersion polymerization.[2] Later, several other methodologies were suggested for the preparation of nanoparticles capable of encapsulating desired biomolecules. Polymeric nanoparticles, more specifically nanospheres, have been explicitly employed for the encapsulation and delivery of lipophilic and hydrophilic drugs.

Polymeric nanoparticles can be fabricated in two different ways: by polymerization of monomers or from preformed polymers.[3] To date, a large number of methodologies have been proposed for the preparation of nanoparticles depending upon the type of molecules to be delivered. Polymeric nanoparticles can be prepared by polymerization of monomers using various polymerization techniques such as emulsion, dispersion, and interfacial polymerization. However, nanoparticles from polymers can be prepared by different approaches including solvent evaporation, salting-out, thermal denaturation, dialysis, and supercritical fluid technology. An illustration of various techniques employed for the preparation of polymeric nanoparticles is given in Figure 2.1. The choice

Published by Woodhead Publishing Limited, 2013

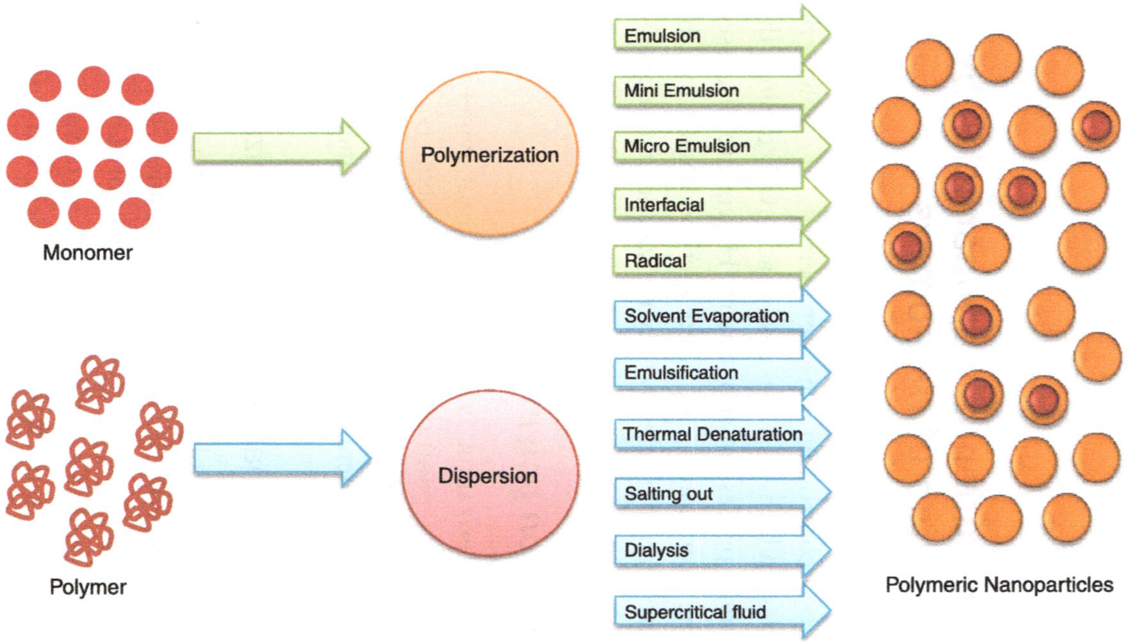

Figure 2.1 An illustration of various techniques for the preparation of polymer nanoparticles

of preparation method depends on a number of parameters such as desired application, the type of polymeric system, size requirement, etc.

2.2 Preparation of nanoparticles by polymerization of monomers

Nanoparticles suitable for specific applications can be prepared by controlling the properties of polymeric nanoparticles, which can be easily achieved during the polymerization process of monomers. Herein are discussed some of the commonly employed methodologies for the preparation of nanoparticles by polymerization of monomers.

2.2.1 Emulsion polymerization

Emulsion polymerization is one of the most widely used methods for the preparation of specialty polymers. On the basis of use of surfactant, it can be sub-divided into conventional and surfactant-free emulsion polymerization.

2.2.1.1 Conventional emulsion polymerization

This is the most commonly employed method of nanoparticle preparation from polymerization of monomers. The key ingredients comprise water, a monomer of low water solubility, water-soluble initiator, and a surfactant. This method consists of emulsification of monomers in an aqueous phase and generation of initiator radicals that can diffuse into the monomer-swollen micelles. The anionic polymerization in the micelles is initiated by water, and proceeds further by additional migration of monomers into the micelles. Chain

Published by Woodhead Publishing Limited, 2013

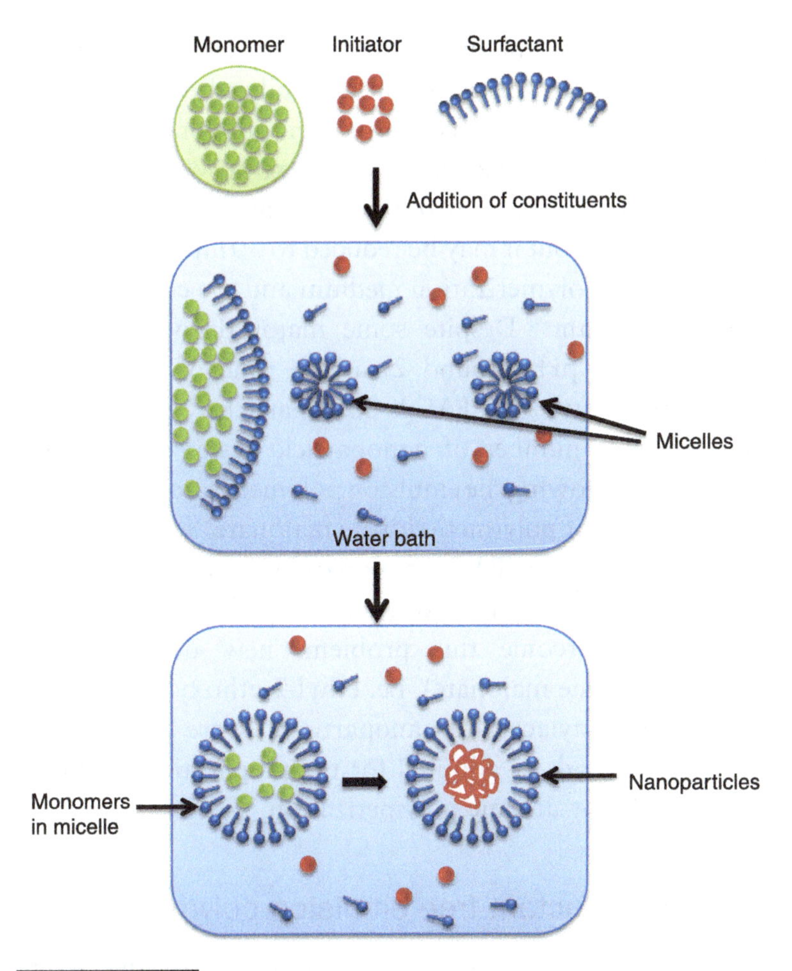

Figure 2.2 Mechanism of emulsion polymerization

transfer agents are abundant, and termination occurs by radical combination (Figure 2.2).

Preparation of poly(alkyl cyanoacrylate) (PACA) nanoparticles by emulsion polymerization was reported by Couvreur et al.[4,5] Anionic polymerization can take place in micelles after the diffusion of monomer molecules through the water phase and is initiated by water itself. Both the rate of polymerization and the absorption of the drug

Published by Woodhead Publishing Limited, 2013

are determined by the pH of the medium when the latter is ionizable. Usually, the emulsion is stabilized either by non-ionic surfactants, such as poloxomer 188 or polysorbate 20, or by steric stabilizers, such as dextran 70. The size of the nanoparticles synthesized by this method is generally around 200 nm, but it may be reduced to 30 nm by increasing the pH of the polymerization medium and the concentration of the surfactant.[6] Despite some major drawbacks, such as use of low pH (around 2) at the time of preparation and cytotoxicity, the PACA nanoparticles gained wide popularity.[7] A number of nanoparticle systems have been synthesized following the emulsion polymerization technique. Nanoparticles of poly(methyldiene malonate) were prepared as a replacement for PACA nanoparticles, but later on they were found to be non-biodegradable both *in vivo* and *in vitro*.[8] To overcome this problem, new derivatives of poly(methyldiene malonate), i.e. ethyl-2-ethoxycarbonylmet hylenoxycarbonylacrylate nanoparticles, were prepared by the same methods as adopted for the preparation of PACA nanoparticles by anionic polymerization.[8]

2.2.1.2 Surfactant free emulsion polymerization

This methodology gained significant popularity due to its simple, green process for preparation of polymeric nanoparticles without the use of stabilizing surfactants and the inconvenience of removing them afterwards.[9–13] This emulsifier free reaction system consists of deionized water, a water-soluble initiator (i.e. potassium persulfate (KPS)), and monomers, such as acryl or vinyl monomers. The stabilization of polymeric nanoparticles in such a process takes place via the use of ionizable initiators or ionic co-monomers. In one study, PMMA nanoparticles were prepared by using this methodology, in which polymerization was stimulated with

Published by Woodhead Publishing Limited, 2013

microwave irradiation.[14] It was reported that the average particle size was primarily controlled by the monomer methyl methacrylate concentration. The particle size increased from 103 nm to 215 nm when the concentration was increased from 0 to 0.3 mol/L. Further, the nanoparticle size could be controlled by using cross-linkers with enhanced reactivity through a one-step microwaving process. The size of the nanoparticles was successfully controlled by limiting the cross-linking to intra-particle cross-linking rather than inter-particle cross-linking.[15] Polyacrylate nanoparticles were prepared by employing sodium salt hydrate (NaSS) as the stabilizing agent, with a particle size of 172.5 nm; a reduction in particle size from 263.4 nm to 172.5 nm was observed with manipulation of NaSS concentration.[16] Polystyrene nanoparticles of particle size 200–250 nm were prepared using ultrasonic irradiation, an anionic ionizable water-soluble initiator, KPS, and cetyl alcohol as the co-stabilizer.[17] Emulsion polymerization has several advantages, but its applications are limited by its disadvantages, such as inability to synthesize, monodisperse and precisely control particle size.[18]

2.2.2 Mini-emulsion polymerization

Application of the mini-emulsion polymerization process for nanoparticle preparation has recently gained momentum. The reagents employed in this process consist of water, monomer mixture, co-stabilizer, surfactant, and initiator (Figure 2.3). Although the reagents used in mini-emulsion polymerization are similar to those of emulsion polymerization, the major difference is the use of a low molecular mass compound as the co-stabilizer and also the utilization of a high shear device (ultrasound, etc.). Mini-emulsions are precariously stabilized, require a high

Published by Woodhead Publishing Limited, 2013

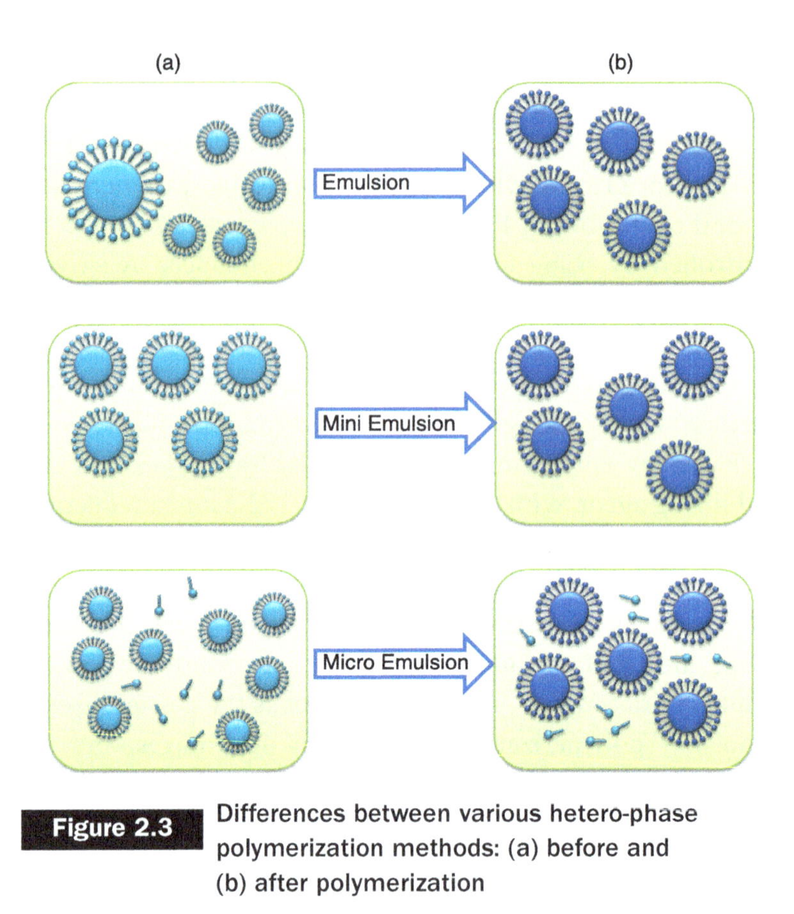

Figure 2.3 Differences between various hetero-phase polymerization methods: (a) before and (b) after polymerization

shear to reach a steady state and have an interfacial tension higher than zero.

PMMA and poly(n-butylacrylate) nanoparticles were prepared using a mini-emulsion polymerization technique by employing sodium lauryl sulfate (SLS)/n-dodecyl mercaptan (DDM) and SLS/hexadecane as surfactant/co-stabilizer systems, respectively.[19] Polyacrylic acid nanoparticles were synthesized using a co-emulsifier system consisting of a mixture of Span 80 and Tween 80.[20] The polymerization was initiated by free radicals, and the particle size was dictated by

Published by Woodhead Publishing Limited, 2013

the type of radical initiator used. Application of water-soluble initiators, such as ammonium persulfate (APS), resulted in microparticles; however, nanoparticles with average diameter between 80 and 150 nm were generated when lipophilic radical initiators, such as azobisisobutyronitrile (AIBN), were used. Polystyrene single-walled carbon nanotube composites were prepared using mini-emulsion polymerization, in which sodium dodecyl sulfate (SDS) and 1-pentanol were employed as the surfactant and co-surfactant, respectively.[21] Later, synthesis of polystyrene and PMMA nanoparticles was reported using Lutensol AT 50 as surfactant.[22]

2.2.3 Micro-emulsion polymerization

Micro-emulsion polymerization is a comparatively new and efficient strategy for the preparation of polymeric nanoparticles. Emulsion polymerization exhibits three reaction rate intervals, whereas only two are detected in micro-emulsion polymerization. Also, the particle size and the average number of chains per particle are comparatively smaller in micro-emulsion polymerization. In micro-emulsion polymerization, a water-soluble initiator is added to the aqueous phase of a thermodynamically stable micro-emulsion containing swollen micelles (Figure 2.3). The polymerization starts from this thermodynamically stable, spontaneously formed state and relies on high quantities of surfactant systems, which possess an interfacial tension at the oil/water interface close to zero. Furthermore, the particles are completely covered with surfactant because of the use of a large amount of surfactant. Initially, polymer chains are formed only in some droplets, as the initiation cannot be attained simultaneously in all micro-droplets.

Published by Woodhead Publishing Limited, 2013

Later, the osmotic and elastic influence of the chains destabilizes the fragile micro-emulsions and leads to an increase in the particle size, the formation of empty micelles, and secondary nucleation.

Polymerization of vinyl acetate in micro-emulsions stabilized with sodium bis(2-ethylhexyl) sulfosuccinate (AOT) was examined as a function of concentration and type of initiator (azobisisobutylnamide dihydrochloride (AIBA) and KPS) and temperature.[23] Nanoparticles smaller than 40 nm in diameter were prepared using micro-emulsion systems containing 1% AOT and 3% VA-044. A comparative study involving preparation of poly(vinylacetate) nanoparticles by emulsion and micro-emulsion polymerization approaches revealed that the poly(vinyl acetate) in the high solid content latex has much smaller molar masses than the poly(vinyl acetate) in emulsion made lattices with similar solid content.[24] Also, the micro-emulsion made lattices contain particles two- to three-fold smaller than those obtained by emulsion polymerization. Polyhexylmethacrylate nanoparticles with a size range of 38–53 nm were prepared using a mixture of dodecyltrimethylammonium bromide (DTAB) and dimethyldioctadecylammonium bromide (DDAB) as stabilizer and VA-044 as the initiator.[25] Another study employed a surfactant mixture of SDS and AOT in micro-emulsion polymerization of butylacrylate, and nanoparticles with a particle size smaller than 40 nm were obtained.[26] Though polymer latexes obtained by micro-emulsion polymerization found numerous applications, their commercial use has been hampered due to very dilute formulations and requirement of a high ratio of surfactant to monomer. In typical preparation processes, the surfactant concentrations usually exceed the amount required for polymer stability.

Published by Woodhead Publishing Limited, 2013

2.2.4 Interfacial polymerization

This technique of nanoparticle preparation has been exclusively used to synthesize nanocapsules. In this method the drug and an organic liquid are combined, then slowly mixed with an aqueous phase with continuous stirring (Figure 2.4). A diffusing co-solvent (oil-soluble and miscible in water) is then used to bring the monomer to the surface of the droplets. Spontaneous emulsification at the organic/aqueous interface leads to polymerization. The resultant nanocapsules with oil-filled cavities are then separated by ultracentrifugation. This method of nanocapsule preparation is a relatively new technique and is more complicated than that of nanospheres. For the interfacial technique to be successful, bulk polymerization should not occur in the droplet. Furthermore, an appropriate phase composition and co-solvent are critical.[27]

Al Khouri Fallouh et al. (1986) developed the technique of interfacial polymerization for alkylcyanoacrylate monomers

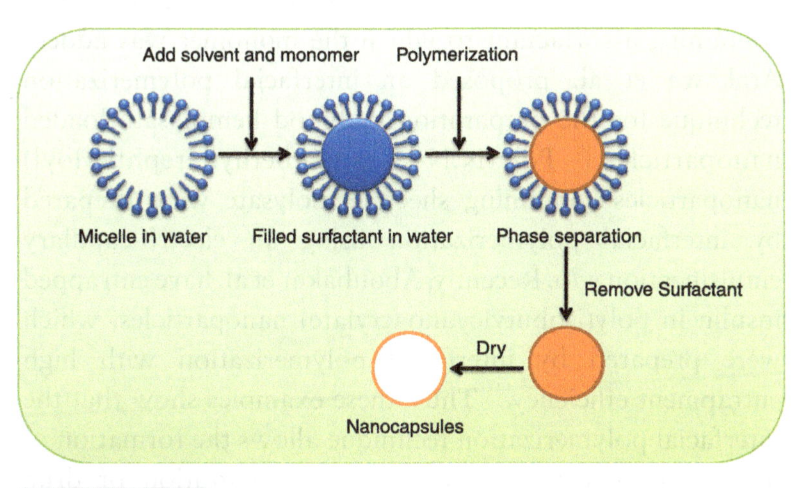

Figure 2.4 Mechanism of interfacial polymerization

Published by Woodhead Publishing Limited, 2013

with the aim of entrapping lipophilic drugs.[28] In this technique, the alkylcyanoacrylate monomers and the drug were dissolved in an ethanolic phase containing an oil, a phospholipid mixture or benzyl benzoate. The phase was then slowly added into water containing a non-ionic surfactant (e.g. poloxamer 188). Dispersion of the organic phase occurred simultaneously with both the diffusion of the ethanol in the aqueous phase and anionic polymerization of the monomer at the water–oil interface. The polymerization was initiated by hydroxide ions and led to the formation of nanocapsules with an oily core and a polymeric shell. The nanocapsule suspension was then concentrated under reduced pressure. Although there was twice as much surfactant as polymer in the raw suspension, no purification was described.

A significant amount of effort has been made to entrap hydrophilic drug in the nanoparticles prepared by interfacial polymerization. El Samaligy et al. proposed water-in-oil (w/o) emulsification followed by an interfacial polymerization method.[29] Thc successful entrapment of doxorubicin was demonstrated by this procedure. The aqueous solution was emulsified in an organic phase (cyclohexane or chloroform) containing a surfactant to which the monomer was added. Arakawa et al. proposed an interfacial polymerization technique for the preparation of blood hemolysate loaded nanoparticles.[30] Poly(N,N-L-lysine-diethylterephthaloyl) nanoparticles containing sheep hemolysate were prepared by interfacial polymerization using an electro-capillary emulsification w/o. Recently, Aboubakar et al. have entrapped insulin in poly(isobutylcyanoacrylate) nanoparticles, which were prepared by interfacial polymerization with high entrapment efficiency.[31] Thus, these examples show that the interfacial polymerization technique allows the formation of nanoparticles with relatively high concentration of drug entrapment.

Published by Woodhead Publishing Limited, 2013

2.2.5 *Radical polymerization*

Recent developments related to controlled radical polymerization methods have opened up new avenues for this process of polymeric nanoparticle preparation.[32,33] Three different approaches have been successfully employed for controlled polymerizations: nitroxide-mediated polymerization (NMP), atom transfer radical polymerization (ATRP), and reversible addition and fragmentation transfer chain polymerization (RAFT).[34–36] The size of the nanoparticles depends on the nature and concentration of the controlling agent, monomer, initiator, and emulsion type (apart from temperature). Among all the factors, the nature of the controlling agent significantly dictates the size of the nanoparticles formed. Polystyrene nanoparticles have been prepared by using nitroxide as a controlling agent, and its effect on the particle size has been investigated.[37] Formation of stable latex nanoparticles with small particle size (below 100 nm) and good consistency was observed with the use of acetoxy derivatives (N6) and SG1 (N8). Matyjaszewski et al. prepared poly(n-butylacrylate) nanoparticles of average size 300 nm, which varied with the concentrations of the initiator, the mediating (control) agent, the surfactant, and the temperature.[38]

2.3 Preparation of nanoparticles using preformed polymers

Nanoparticles can also be prepared from macromolecules (natural and synthetic) such as polymers, proteins and polysaccharides. The major advantage of this method is getting rid of the presence of potentially toxic residual monomers, oligomers, surfactant, initiators, etc. in the nanoparticles and to avoid further tedious purification of the nanoparticles. To date several methods have been reported for the preparation of biodegradable nanoparticles using preformed polymers such as poly(lactic acid) (PLA), poly(l-glycolide) (PLG), poly(lactide-co-glycolide) (PLGA), and PACA.[39]

2.3.1 Solvent evaporation

Solvent evaporation was the first method developed for the preparation of nanoparticles from preformed polymers. The traditional method involves use of two different approaches for the emulsion formation: the preparation of single emulsions, e.g. oil-in-water (o/w), or double emulsions, e.g. (water-in-oil)-in-water, (w/o)/w. In single emulsion methodology, the polymer is suspended in a volatile and water immiscible solvent like dichloromethane, chloroform, ethyl acetate, etc. The drug molecule to be entrapped is dissolved into the preformed polymer solution and this organic phase is dispersed in the aqueous phase using conventional emulsification techniques and surfactant/emulsifiers like gelatin, poly(vinyl alcohol) (PVA), polysorbate-80, poloxamer 188, etc. The polymeric o/w droplets thus formed are separated by the removal of the organic solvent following distillation under reduced pressure. Afterwards, the solidified nanoparticles can be collected by ultracentrifugation and washed with distilled water to remove surfactants, followed by lyophilization. One of the major drawbacks of this methodology is the formation of highly polydispersed particles, though the size is smaller than 1 μm.

PLGA nanoparticles with particle size range of 60–200 nm were prepared by employing dichloromethane and acetone (8:2, v/v) as the solvent system and PVA as the stabilizing

Published by Woodhead Publishing Limited, 2013

agent.[40] Later, PLGA nanoparticles of average size 200 nm were produced by Lemoine et al. by using dichloromethane 1.0% (w/v) as the solvent and PVA or Span 40 as the stabilizing agent.[41] Both the single emulsion and double emulsion methods were employed for the preparation of nanoparticles, and the effect of the emulsion method and the experimental parameters on the particle size were also investigated. In the solvent evaporation method, the mixing technique plays a significant role in the preparation of nanoparticles. Bilati et al. investigated the effect of the sonication process on the characteristics of PLGA nanoparticles prepared by the w/o/w solvent evaporation method.[42] The duration and intensity of sonication were investigated with respect to the ability to modify the size and distribution of the nanoparticles. It was demonstrated that the duration of the second mixing step, which leads to the w/o/w emulsion, has a greater influence on the final mean particle size than the first step for the water-in-oil emulsion. When the second emulsification time period was increased, the mean particle size decreased to a certain level. This study suggested that a threshold exists for the sonication intensity leading to a controlled particle size with a narrow distribution.

2.3.2 Emulsification/solvent diffusion method

This methodology is derived from the solvent evaporation method, in which a water-soluble solvent like acetone or methanol is used along with a water-insoluble organic solvent like dichloromethane or chloroform.[43] Due to the spontaneous diffusion of water-soluble solvent (acetone or methanol), an interfacial turbulence is created between two phases leading to the formation of smaller particles. As the concentration of the water-soluble solvent (acetone)

Published by Woodhead Publishing Limited, 2013

increases, a considerable decrease in particle size can be achieved. El-Shabouri reported preparation of nanoparticles by the emulsification solvent diffusion method, using lecithin and poloxamer 188 as emulsifiers, and chitosan HCl, gelatin-A or sodium glycocholate (SGC) as charge inducing agents.[44] This method is based on the partial miscibility of an organic solvent with water. An o/w emulsion is obtained upon injection of organic phase into chitosan solution containing a stabilizing agent (i.e. poloxamer) under mechanical stirring, followed by high pressure homogenization. The emulsion is then diluted with a large amount of water to overcome organic solvent miscibility in water. Polymer precipitation occurs as a result of the diffusion of organic solvent into water, leading to the formation of nanoparticles. This method is suitable for hydrophobic drugs and showed a high percentage of drug entrapment. The major drawbacks of this method include harsh processing conditions (e.g. the use of organic solvents) and the high shear force that is used during nanoparticle preparation.

2.3.3 Thermal denaturation

Preparation of nanoparticles by thermal denaturation was first reported by Scheffel et al. from a natural macromolecule, albumin.[45] Heat treatment of albumin is applicable only to drug molecules that are not heat sensitive. Initially, the albumin microspheres of size 12–45 μm were prepared by emulsification of the aqueous solution of albumin in cotton seed oil, followed by heating for total evaporation of the water along with the denaturation of albumin.[46] This technique was further adapted by Scheffel et al. to prepare an emulsion (aqueous solution of albumin in cotton seed oil) characterized by very small droplets in the dispersed phase. The emulsion was then added to cotton seed oil pre-heated

Published by Woodhead Publishing Limited, 2013

to above 100°C. The addition of the emulsion to the hot oil instantaneously leads to the evaporation of the water contained in the droplets as well as denaturation of the albumin. Later, Kramer used this method for the preparation of drug loaded nanoparticles for sustained drug delivery studies.[47] Subsequently, Widder et al. also used this method for the preparation of magnetized albumin nanoparticles.[48]

2.3.4 Salting-out

The salting-out method for nanoparticles preparation is based on the separation of a water miscible solvent from aqueous solution via a salting-out effect. The salting-out approach can be considered as a modification of the emulsification/solvent diffusion method. The initial step consists of dissolving polymer and drug in a solvent such as acetone, followed by subsequent emulsification into an aqueous gel which contains the salting-out agent (electrolytes, such as magnesium chloride, calcium chloride, and magnesium acetate, or non-electrolytes such as sucrose) and a colloidal stabilizer such as polyvinylpyrrolidone (PVP) or hydroxyl ethylcellulose. This oil/water emulsion is diluted with a sufficient volume of water or aqueous solution to enhance the diffusion of acetone into the aqueous phase, thus resulting in the formation of nanospheres. The choice of the salting-out agent is critical, as it can play an important role in the encapsulation efficiency of the drug. At the final step, both the solvent and the salting-out agent are then removed by cross-flow filtration.[49] Leroux et al. reported preparation of PLA nanoparticles by adding an aqueous gel containing magnesium acetate tetrahydrate and PVA to an acetone solution of the polymer, forming a water-in-oil emulsion.[50] Although water was miscible with acetone, a

Published by Woodhead Publishing Limited, 2013

liquid–liquid two-phase system was formed due to the presence of the salting-out agent. Further addition of the aqueous gel resulted in an oil-in-water emulsion. Finally, a sufficient amount of water was added to allow diffusion of acetone into the aqueous phase, which produced nanoparticles with an average size of 295 nm.

2.3.5 Dialysis

Dialysis is a simple and efficient method for the preparation of polymeric nanoparticles with small size range and narrow distribution. In this method, the desired polymer is dissolved in an organic solvent and placed inside dialysis tubing with a suitable molecular weight (MW) cut-off (Figure 2.5). Dialysis is usually performed against a non-miscible solvent. The removal of the solvent from the dialysis tubing is followed by gradual aggregation of polymer due to loss of solubility and fabrication of a homogeneous suspension of nanoparticles. Nanoparticles of poly(γ-glutamic acid) were prepared using different solvents such as dimethylsulfoxide (DMSO), dimethylformamide (DMF), dimethylacetamide (DMAc), and N-methyl-2-pyrrolidone (NMPy); spherical nanoparticles with diameters in the size

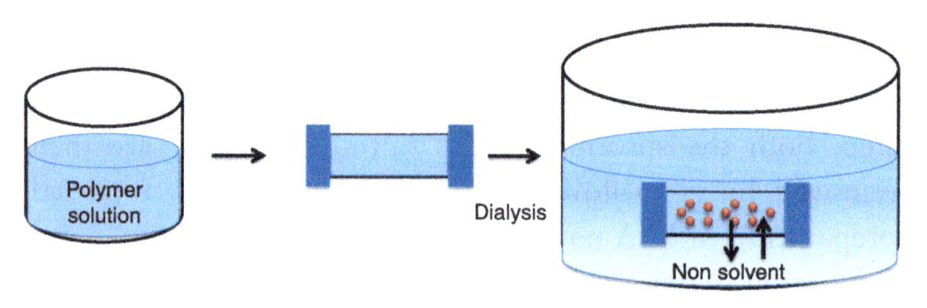

Figure 2.5 Schematic representation of dialysis method for preparation of polymer nanoparticles

Published by Woodhead Publishing Limited, 2013

range of 100 to 200 nm were observed in the case of DMF.[51] In contrast, use of NMPy solvent resulted in formation of various sizes of nanoparticles with larger size distribution. Poly(d,l-lactide)-g-poly(N-isopropyl acrylamide-co-methacrylic acid) (PLA-g-P(NIPAm-co-MAA)) graft co-polymers have been employed for the preparation of "intelligent" nanoparticles with multi-functional sensitivity using the dialysis method with DMSO as the solvent.[52] In another study, poly(lactic acid) branched polyethylene glycol (PLA-b-PEG) nanoparticles of spherical shape with a size range of 90–330 nm were synthesized by employing DMSO as the solvent.[53]

In one of our studies, PEI nanoparticles were prepared by cross-linking with PEG-bis-(p-nitrophenyl carbonate) with a size range of 18–75 nm (hydrodynamic radii) with almost uniform population.[54] The p-nitrophenoxide ions generated during the reaction were easily removed by dialysis against a 5% sodium bicarbonate solution for 3 days prior to dialyzing against distilled water.

2.3.6 Supercritical fluid technology

Supercritical or compressed fluid has been successfully employed for the preparation of nanoparticles and microparticles from polymers.[55] This method involves solubilization of polymer in a supercritical fluid, followed by expansion of the solution through a nozzle. The supercritical fluid gets evaporated during the spraying process, resulting in precipitation of the solute particles. This method yields pure nanoparticles as the precipitated solute is free from any solvent content. It has other advantages, such as suitable technological and biopharmaceutical properties, along with high quality. It has been successfully implicated in several applications involving protein drug delivery systems. PEG/

PLA nanoparticles encapsulating insulin were prepared employing this technique.[56]

Meziani et al. used the rapid expansion of a supercritical solution into a liquid solvent (RESOLV) technique for the preparation of poly(heptadecafluorodecylacrylate) (PHDFDA) nanoparticles having an average size of less than 50 nm.[57] The polymer was soluble in supercritical CO_2, but insoluble in water. In a typical experiment, a CO_2 solution of the polymer PHDFDA (0.3 wt. %) was pressurized in a syringe pump and pushed through the heating unit to reach the desired supercritical temperature before reaching the expansion nozzle. The expanding solution passed through the nozzle into a chamber containing water at ambient temperature. Since the polymer was insoluble in water, it precipitated to form nanoparticles. The expansion of the supercritical CO_2 solution of 60 ml took about 30 min. In the first 5 min of the rapid expansion, the aqueous nanoparticle suspension appeared clear and stable. However, scanning electron microscope (SEM) micrographs revealed that, as the rapid expansion progressed, larger particles were formed in the aqueous suspension due to aggregation of the initially formed nanoparticles. The use of an aqueous NaCl solution in place of water at the receiving end of the expansion process stabilized the initially formed nanoparticles to some extent due to the increased ionic strength in the suspension.

2.4 Methods of controlled release

Two types of controlled drug release can be achieved – temporal and distribution control. In temporal control, the aim of the drug delivery system is to deliver the drug over an extended period of time or at a specific duration during treatment. This type of control is highly beneficial for drugs

Published by Woodhead Publishing Limited, 2013

having fast metabolism and elimination after administration. In distribution control, the aim of the drug delivery system is to target the release of the drug to the precise site of activity within the body. This type of control is beneficial for two principal situations: (i) when the natural distribution causes drug molecules to encounter tissues and cause major side effects, thus prohibiting further treatment, and (ii) when the natural distribution of the drug does not allow the drug molecule to reach its molecular site of action. A wide variety of drugs, including chemotherapeutic drugs, immune suppressants, antibiotics, steroids, hormones, etc., have benefited from this temporal and distribution controlled release of drugs.

In controlled release, the physical and chemical properties of the system are modified to achieve the drug release kinetics. The main factors involved in achieving the desired releasc rates are:

- desorption of the surface bound/adsorbed drug;
- diffusion through the nanoparticle matrix;
- diffusion through the polymer wall (in the case of nanocapsules);
- nanoparticle matrix erosion;
- a combined erosion/diffusion process.

Diffusion and biodegradation are the major governing factors for the process of drug release. Drug release profiles from nanoparticles depend upon the nature of the delivery system. Such systems often use synthetic polymers as carriers for the drug. In the case of a matrix device, drug is uniformly distributed/dissolved in the matrix and the release occurs by the diffusion or erosion of the matrix. If the diffusion of the drug is faster than matrix degradation, then the mechanism of drug release occurs mainly by diffusion; otherwise it depends upon degradation of the matrix.[58] Rapid initial release is attributed to the fraction of the drug which is adsorbed or weakly bound to the large surface area of the nanoparticles, rather than to the drug incorporated in the nanoparticles.

Diffusional release methods can be categorized as reservoir devices, monolithic devices and hollow fibers. These devices exhibit the classical biphasic Michaelis–Menten kinetics in which the initial burst phase is followed by a steady release rate. Recently, Polakovic et al. theoretically studied the release from PLA nanoparticles with varying amounts (7–32% w/w) of lidocaine.[59] In the case of nanocapsules, the drug core is coated with the polymer and release occurs by diffusion of the drugs from the core across the polymeric barrier layer. Hence, theoretically the drug release should follow zero order kinetics. Cavallaro et al. obtained similar release profiles for indomethacin from bulk nanoparticles and nanocapsules, thus indicating the non-interfering nature of the polymer coating, so that release is caused by partitioning of the drug.[60]

In monolithic devices, the therapeutic agent is dissolved or dispersed at a rate controlled by the polymeric matrix. These devices do not exhibit zero order kinetics, but are the simplest and the most convenient for applications with the prolonged release of an active agent. Fresta et al. reported a higher burst, up to 60–70%, for the nanoparticles loaded with the drug by absorption. Here the burst effect is less and the remaining drug release is quite slow.[61]

In osmosis the drug is released from the semi-permeable membrane. This type of delivery is independent of the agent's properties, has a constant rate of delivery and exhibits greater rates than diffusion control. Osmotic systems are pH dependent and are very useful in gastrointestinal drug delivery. Polymer bio-erosion is the conversion of an initially

Published by Woodhead Publishing Limited, 2013

water-insoluble system to a water-soluble one. This is often classified into three types, depending upon which portion of the drug is cleaved (cross-links, pendant groups or backbone). In the case of drug release from hydrogel nanoparticles, release occurs mainly due to swelling, which can be controlled either by adding hydrophilic functional groups or by monitoring the cross-linking of the matrix.

It is evident that the various types of controlled release nanoparticulate systems have the potential to have a revolutionary impact on drug delivery. The recent advances in the area of peptide design and gene therapy have necessitated the need to develop newer controlled release systems.

2.5 References

1. Birrenbach, G. and Speiser, P.P. (1976) Polymerized micelles and their use as adjuvants in immunology. *J Pharm Sci*, **65**, 1763–6.
2. Kreuter, J. and Speiser, P.P. (1976) In vitro studies of poly(methyl methacrylate) adjuvants. *J Pharm Sci*, **65**, 1624–7.
3. Soppimath, K.S., Aminabhavi, T.M., Kulkarni, A.R. and Rudzinski, W.E. (2001) Biodegradable polymeric nanoparticles as drug delivery devices. *J Control Release*, **70**, 1–20.
4. Couvreur, P., Kante, B., Roland, M., Guiot, P., Bauduin, P. et al. (1979) Polycyanoacrylate nanocapsules as potential lysosomotropic carriers: preparation, morphological and sorptive properties. *J Pharm Pharmacol*, **31**, 331–2.
5. Couvreur, P., Kante, B. and Roland, M. (1978) Perspective on the use of microdisperse forms as intracellular vehicles. *Pharm Acta Helv*, **53**, 341–7.

Published by Woodhead Publishing Limited, 2013

6. Seijo, B., Fattal, E., Roblot-Treupel, L. and Couvreur, P. (1990) Design of nanoparticles of less than 50 nm diameter: preparation, characterization and drug loading. *Int J Pharm*, **62**, 1–7.
7. Lherm, C., Müller, R.H., Puisieux, F. and Couvreur, P. (1992) Alkylcyanoacrylate drug carriers: II. Cytotoxicity of cyanoacrylate nanoparticles with different alkyl chain length. *Int J Pharm*, **84**, 13–22.
8. De Keyser, J.-L., De Cock, C.J.C., Poupaert, J.H. and Dumont, P. (1989) Synthesis of 14C labelled acrylic derivatives: Diethyl [3–14C] methylidenemalonate and isobutyl [3–14C] cyanoacrylate. *J Labelled Comp Radiopharm*, **27**, 909–16.
9. Hearn, J., Wilkinson, M.C., Goodall, A.R. and Chainey M. (1985) Kinetics of the surfactant-free emulsion polymerization of styrene: the post nucleation stage. *J Polym Sci Polym Chem Ed*, **23**, 1869–83.
10. Song, Z. and Poehelin, G.W. (1990) Kinetics of emulsifier-frcc emulsion polymerization of styrene. *J Polym Sci Part A Polym Chem*, **28**, 2359–92.
11. Zou, D., Ma, S., Guan, R., Park, M., Sun, L. et al. (1992) Model filled polymers V. Synthesis of crosslinked monodisperse polymethacrylate beads. *J Polym Sci Part A Polym Chem*, **30**, 137–44.
12. Shouldice, G.T.D., Vandezande, G.A. and Rudin, A. (1994) Practical aspects of the emulsifier-free emulsion polymerization of styrene. *Eur Polym J*, **30**, 179–83.
13. Pang, S.W., Park, H.Y., Jang, Y.S., Kim W.S. and Kim, J.H. (2002) Effects of charge density and particle size of poly[styrene/(dimethylamino)ethyl methacrylate] nanoparticle for gene delivery in 293 cells. *Colloid Surf B*, **26**, 213–22.

Published by Woodhead Publishing Limited, 2013

14. Bao, J. and Zhang, A. (2004) Poly(methyl methacrylate) nanoparticles prepared through microwave emulsion polymerization. *J Appl Polym Sci*, **93**, 2815–20.
15. An, Z., Tang, W., Hawker, C.J. and Stucky, G.D. (2006) One-step microwave preparation of well-defined and functionalized polymeric nanoparticles. *J Am Chem Soc*, **128**, 15054–5.
16. Cui, X., Zhong, S. and Wang, H. (2007) Emulsifier-free core–shell polyacrylate latex nanoparticles containing fluorine and silicon in shell. *Polymer*, **48**, 7241–8.
17. Faridi-Majidi, R. and Sharifi-Sanjani, N. (2007) Emulsifier-free miniemulsion polymerization of styrene and the investigation of encapsulation of nanoparticles with polystyrene via this procedure using an anionic initiator. *J Appl Polym Sci*, **105**, 1244–50.
18. Zhang, G., Niu, A., Peng, S., Jiang, M., Tu, Y. et al. (2001) Formation of novel polymeric nanoparticles. *Acc Chem Res*, **34**, 249–56.
19. Leiza, J.R., Sudol, E.D. and El-Aasser, M.S. (1997) Preparation of high solids content poly(n-butyl acrylate) latexes through miniemulsion polymerization. *J Appl Polym Sci*, **64**, 1797–809.
20. Kriwet, B., Walter, E. and Kissel, T. (1998) Synthesis of bioadhesive poly(acrylic acid) nano- and microparticles using an inverse emulsion polymerization method for the entrapment of hydrophilic drug candidates. *J Control Release*, **56**, 149–58.
21. Ham, H.T., Choi, Y.S., Chee, M.G. and Chung, I.J. (2006) Singlewall carbon nanotubes covered with polystyrene nanoparticles by in-situ miniemulsion polymerization. *J Polym Sci A Polym Chem*, **44**, 573–84.
22. Ziegler, A., Landfester, K. and Musyanovych, A. (2009) Synthesis of phosphonate-functionalized polystyrene and poly(methyl methacrylate) particles and their kinetic

Published by Woodhead Publishing Limited, 2013

behavior in miniemulsion polymerization. *Colloid Polym Sci*, **287**, 1261–71.

23. Sosa, N., Zaragoza, E.A., López, R.G., Peralta, R.D., Katime, I. et al. (2000) Unusual Free Radical Polymerization of Vinyl Acetate in Anionic Microemulsion Media. *Langmuir*, **16**, 3612–19.
24. Sosa, N., Peralta, R.D., López, R.G., Ramos, L.F., Katime et al. (2001) A comparison of the characteristics of poly(vinyl acetate) latex with high solid content made by emulsion and semi-continuous microemulsion polymerization. *Polymer*, **42**, 6923–8.
25. Hermanson, K.D. and Kaler, E.W. (2003) Kinetics and mechanism of the multiple addition microemulsion polymerization of hexyl methacrylate. *Macromolecules*, **36**, 1836–42.
26. Ramírez, A.G., López, R.G. and Tauer, K. (2004) Studies on semibatch microemulsion polymerization of butyl acrylate: influence of the potassium peroxodisulfate concentration. *Macromolecules*, **37**, 2738–47.
27. Pierre, G. and Couvreur, P. (1986) In *Polymeric Nanoparticles and Microspheres*. CRC Press, Inc. Florida, USA.
28. Al Khouri Fallouh, N., Roblot-Treupel, L., Fessi, H., Devissaguet, J.P. and Puisieux, F. (1986) Development of a new process for the manufacture of polyisobutylcyanoacrylate nanocapsules. *Int J Pharm*, **28**, 125–32.
29. el-Samaligy, M.S., Rohdewald, P. and Mahmoud, H.A. (1986) Polyalkyl cyanoacrylate nanocapsules. *J Pharm Pharmacol*, **38**, 216–18.
30. Arakawa, M. and Kondo, T. (1981) Preparation of hemolysate-loaded poly(N alpha, N epsilon-L-lysine-diylterephthaloyl) nanocapsules. *J Pharm Sci*, **70**, 354–7.

31. Aboubakar, M., Puisieux, F., Couvreur, P., Deyme, M. and Vauthier, C. (1999) Study of the mechanism of insulin encapsulation in poly(isobutylcyanoacrylate) nanocapsules obtained by interfacial polymerization. *J Biomed Mater Res*, **47**, 568–76.
32. Matyjaszewski, K. and Xia, J. (2001) Atom transfer radical polymerization. *Chem Rev*, **101**, 2921–90.
33. Zetterlund, P.B., Kagawa, Y. and Okubo, M. (2008) Controlled/living radical polymerization in dispersed systems. *Chem Rev*, **108**, 3747–94.
34. Nicolas, J., Ruzette, A.-V., Farcet, C., Gérard, P., Magnet, S. et al. (2007) Nanostructured latex particles synthesized by nitroxide-mediated controlled/living free-radical polymerization in emulsion. *Polymer*, **48**, 7029–40.
35. Li, W. and Matyjaszewski, K. (2009) Star polymers via cross-linking amphiphilic macroinitiators by AGET ATRP in aqueous media. *J Am Chem Soc*, **131**, 10378–9.
36. Rieger, J., Zhang, W., Stoffelbach, F.O. and Charleux, B. (2010) Surfactant-free RAFT emulsion polymerization using poly(N,N-dimethylacrylamide) trithiocarbonate macromolecular chain transfer agents. *Macromolecules*, **43**, 6302–10.
37. Cao, J., He, J., Li, C. and Yang, Y. (2001) Nitroxide-mediated radical polymerization of styrene in emulsion. *Polym J*, **33**, 75–80.
38. Matyjaszewski, K., Qiu, J., Tsarevsky, N.V. and Charleux, B. (2000) Atom transfer radical polymerization of n-butyl methacrylate in an aqueous dispersed system: A miniemulsion approach. *J Polym Sci A Polym Chem*, **38**, 4724–34.
39. Allemann, E., Leroux, J.C., Gurny, R. and Doelker, E. (1993) In vitro extended-release properties of

drug-loaded poly(DL-lactic acid) nanoparticles produced by a salting-out procedure. *Pharm Res*, **10**, 1732–7.

40. Song, C.X., Labhasetwar, V., Murphy, H., Qu, X., Humphrey, W.R. et al. (1997) Formulation and characterization of biodegradable nanoparticles for intravascular local drug delivery. *J Control Release*, **43**, 197–212.
41. Lemoine, D. and Preat, V. (1998) Polymeric nanoparticles as delivery system for influenza virus glycoproteins. *J Control Release*, **54**, 15–27.
42. Bilati, U., Allemann, E. and Doelker, E. (2003) Sonication parameters for the preparation of biodegradable nanocapsules of controlled size by the double emulsion method. *Pharm Dev Technol*, **8**, 1–9.
43. Niwa, T., Takeuchi, H., Hino, T., Kunou, N. and Kawashima, Y. (1993) Preparations of biodegradable nanospheres of water-soluble and insoluble drugs with D,L-lactide/glycolide copolymer by a novel spontaneous emulsification solvent diffusion method, and the drug release behavior. *J Control Release*, **25**, 89–98.
44. El-Shabouri, M.H. (2002) Positively charged nanoparticles for improving the oral bioavailability of cyclosporin-A. *Int J Pharm*, **249**, 101–8.
45. Scheffel, U., Rhodes, B.A., Natarajan, T.K. and Wagner, H.N., Jr. (1972) Albumin microspheres for study of the reticuloendothelial system. *J Nucl Med*, **13**, 498–503.
46. Zolle, I., Rhodes, B.A. and Wagner, H.N., Jr. (1970) Preparation of metabolizable radioactive human serum albumin microspheres for studies of the circulation. *Int J Appl Radiat Isot*, **21**, 155–67.
47. Kramer, P.A. (1974) Letter: Albumin microspheres as vehicles for achieving specificity in drug delivery. *J Pharm Sci*, **63**, 1646–7.

Published by Woodhead Publishing Limited, 2013

48. Widder, K., Flouret, G. and Senyei, A. (1979) Magnetic microspheres: synthesis of a novel parenteral drug carrier. *J Pharm Sci*, **68**, 79–82.
49. Quintanar-Guerrero, D., Allemann, E., Fessi, H. and Doelker, E. (1998) Preparation techniques and mechanisms of formation of biodegradable nanoparticles from preformed polymers. *Drug Dev Ind Pharm*, **24**, 1113–28.
50. Leroux, J.-C., Allémann, E., De Jaeghere, F., Doelker, E. and Gurny, R. (1996) Biodegradable nanoparticles — From sustained release formulations to improved site specific drug delivery. *J Control Release*, **39**, 339–50.
51. Akagi, T., Kaneko, T., Kida, T. and Akashi, M. (2005) Preparation and characterization of biodegradable nanoparticles based on poly(gamma-glutamic acid) with l-phenylalanine as a protein carrier. *J Control Release*, **108**, 226–36.
52. Lo, C.-L., Lin, K.-M. and Hsiue, G.-H. (2005) Preparation and characterization of intelligent core-shell nanoparticles based on poly(d,l-lactide)-g-poly(N-isopropyl acrylamide-co-methacrylic acid). *J Control Release*, **104**, 477–88.
53. Na, K., Lee, K.H., Lee, D.H. and Bae, Y.H. (2006) Biodegradable thermo-sensitive nanoparticles from poly(L-lactic acid)/poly(ethylene glycol) alternating multi-block copolymer for potential anti-cancer drug carrier. *Eur J Pharm Sci*, **27**, 115–22.
54. Nimesh, S., Goyal, A., Pawar, V., Jayaraman, S., Kumar, P. et al. (2006) Polyethylenimine nanoparticles as efficient transfecting agents for mammalian cells. *J Control Release*, **110**, 457–68.
55. Wang, Y., Dave, R.N. and Pfeffer, R. (2004) Polymer coating/encapsulation of nanoparticles using a supercritical anti-solvent process. *J Supercrit Fluids*, **28**, 85–99.

Published by Woodhead Publishing Limited, 2013

56. Elvassore, N., Bertucco, A. and Caliceti, P. (2001) Production of insulin-loaded poly(ethylene glycol)/poly(l-lactide) (PEG/PLA) nanoparticles by gas antisolvent techniques. *J Pharm Sci*, **90**, 1628–36.
57. Meziani, M.J., Pathak, P., Hurezeanu, R., Thies, M.C., Enick, R.M. et al. (2004) Supercritical-fluid processing technique for nanoscale polymer particles. *Angew Chem Int Ed Engl*, **43**, 704–7.
58. Magenheim, B., Levy, M.Y. and Benita, S. (1993) A new in vitro technique for the evaluation of drug release profile from colloidal carriers – ultrafiltration technique at low pressure. *Int J Pharm*, **94**, 115–23.
59. Polakovic, M., Gorner, T., Gref, R. and Dellacherie, E. (1999) Lidocaine loaded biodegradable nanospheres. II. Modelling of drug release. *J Control Release*, **60**, 169–77.
60. Cavallaro, G., Fresta, M., Giammona, G., Puglisi, G. and Villari, A. (1994) Entrapment of β-lactams antibiotics in polyethylcyanoacrylate nanoparticles: Studies on the possible in vivo application of this colloidal delivery system. *Int J Pharm*, **111**, 31–41.
61. Fresta, M., Puglisi, G., Giammona, G., Cavallaro, G., Micali, N. et al. (1995) Pefloxacine mesilate- and ofloxacin-loaded polyethylcyanoacrylate nanoparticles: characterization of the colloidal drug carrier formulation. *J Pharm Sci*, **84**, 895–902.

Published by Woodhead Publishing Limited, 2013

3

Tools and techniques for physico-chemical characterization of nanoparticles

DOI: 10.1533/9781908818645.43

Abstract: The transfection efficiency of nanoparticles is significantly influenced by the particle size, shape and charge. A multitude of techniques, such as TEM and DLS, are required to characterize the physico-chemical properties of nanoparticles which may be relevant to their therapeutic potential. Laser diffraction which works on the Mie and Fraunhofer scattering theory has also been employed for particle size measurements. Further, TEM, AFM and SEM have been used for determination of surface morphology of nanoparticles and laser Doppler velocimetry for surface charge. This chapter accounts for various techniques employed for physico-chemical characterization of nanoparticles.

Key words: size, zeta potential, dynamic light scattering, microscopy, TEM, SEM, AFM, laser diffraction.

Published by Woodhead Publishing Limited, 2013

3.1 Introduction

The therapeutic efficacy of nanoparticles has been well correlated with their physico-chemical properties, namely size, shape, surface chemistry, etc. Numerous studies have reported the correlation between the size and shape of nanoparticles and their drug and gene delivery efficacy.[1,2] Though the physico-chemical characterization of nanoparticles has been well associated with biological interactions and uptake by cells, such measurements are not reliable for predicting the efficacy and toxicity of the nanoparticles.

3.2 Physico-chemical characterization

Several natural and/or synthetic polymers have been employed in preparation of polymeric nanoparticles for therapeutic applications, such as PEI, chitosan, hyaluronic acid (HA), dextran, PEG, *N*-(2-hydroxypropyl) methacrylamide copolymers, PVP and poly(aspartic acid).[3] Various techniques are employed for routine characterization of polymeric nanoparticles; some of these techniques are discussed in detail below.

3.2.1 Size and size distribution

The transfection efficiency of nanoparticles is significantly influenced by the particle size. Hence, most of the formulations, such as DNA/polymer, lipid complexes and liposomes, are prepared with special attention to the particle size.[4–6] Numerous publications have revealed that the particle size significantly dictates their cellular and tissue uptake; in

Published by Woodhead Publishing Limited, 2013

some cell lines, only the sub-micron-sized particles are taken up efficiently but not the larger-sized microparticles (e.g. Hepa 1-6, HepG2, and KLN 205).[7,8] Prabha et al. reported that the smaller-sized nanoparticles (mean diameter 70 nm) showed a 27-fold higher transfection than the larger-sized nanoparticles (mean diameter 202 nm) in COS-7 cell line and a four-fold higher transfection in HEK-293 cell line.[9] In a study by Chithrani et al., the uptake of three different gold nanoparticles with size 14, 50, and 74 nm was investigated in HeLa cells.[10,11] It was observed that the kinetics of uptake as well as the saturation concentration varied with different-sized nanoparticles, with 50 nm particles being the most efficient in their uptake, indicating that there might be an optimal size for efficient nanomaterial uptake into cells. The effect of nanoparticle shape on its internalization was also examined: spherical particles of similar size were taken up 500% more than rod-shaped particles, which is explained by the greater membrane wrapping time required for the elongated particles. Studies done for the determination of nanoparticle size are often complicated by the polydispersity of samples. However, characterization of nanoparticles employing multiple techniques, such as transmission electron microscopy (TEM) and dynamic light scattering (DLS), provides information about size of nanoparticles which may be relevant to their therapeutic potential.

3.2.1.1 Microscopy

Due to the very small size of nanoparticles, they are often not resolvable by optical microscopy; hence higher-resolution techniques such as electron microscopy are required for determination of their size and shape. TEM employs a beam of electrons which is transmitted through an ultra-thin specimen, interacting with the specimen as it passes through.

Published by Woodhead Publishing Limited, 2013

This interaction of electrons forms an image which is further magnified and can be focused onto an imaging device, such as a fluorescent screen or charged coupled device (CCD) camera. Samples for nanoparticle characterization can be either prepared by simple deposition of dilute suspensions on copper grids or fixated using a negative staining material such as uranyl acetate. TEM provides information not only about the size of nanoparticles but also about the surface morphology. In one of our studies, samples for TEM were prepared by dispersing lyophilized powder (2 mg) of nanoparticles by sonication in distilled water (1 ml) to obtain a clear suspension.[12] The sample solution (3 μl) was put on a formvar (polyvinyl formal) coated copper grid and air-dried prior to analysis. TEM analysis of cross-linked acrylamido-2-deoxy-glucose (AADG) nanoparticle images showed spherical and compact particles with average size of 85 nm (Figure 3.1). In another study, prior to visualization of samples, the grids were negatively stained with a saturated solution of uranyl acetate.[13]

Atomic force microscopy (AFM) or scanning force microscopy (SFM) is a high-resolution scanning probe microscope with resolution in the range of few nanometers. The instrumental set-up of AFM consists of a cantilever with a very fine tip (probe) at its end, typically silicon with a radius of few nanometers, which is employed to scan the sample surface. At the time of scanning, the forces between the cantilever tip and the sample lead to deflection of the cantilever, which can be detected by using a laser spot reflected from the top surface of the cantilever. AFM is preferred for characterization of nanoparticles as it possesses 3D visualization capability and provides both qualitative and quantitative information about the sample topology, including morphology, surface texture and roughness, and, more important in this case, the size of the particles.[14]

Published by Woodhead Publishing Limited, 2013

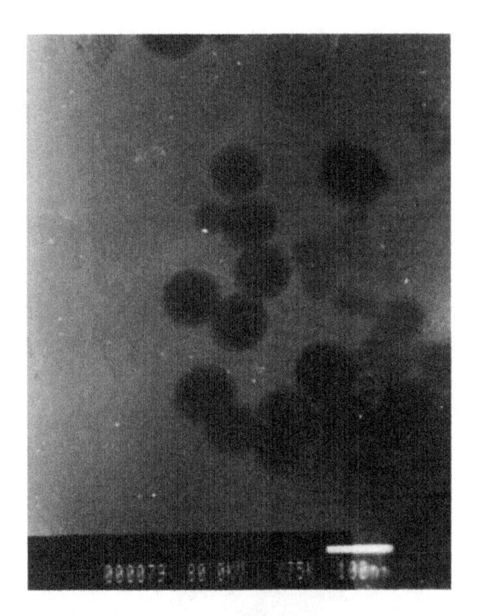

Figure 3.1 **TEM image of AADG nanoparticles. Average size of nanoparticles is 85 nm**

Source: adapted from Nimesh et al. 2006[12]

Recently, AFM has been employed for the estimation of size and surface morphology of nanoparticles. In one of our studies, lyophilized powder (~0.5 mg) of nanoparticles was dispersed by sonication in double distilled water (1 ml) to obtain a suspension, and 2–3 μl of this suspension was deposited on a "Piranha" cleaned glass slide and allowed to dry overnight at room temperature.[15] Subsequently, the glass surface containing the nanoparticles was imaged by AFM. The average size of nanoparticles PPA3 and PPA3 loaded with DNA was found to be 100 nm (Figure 3.2). In another study, the size of guanidinated-PEI/pDNA complexes, prepared at N/P ratio of 30 in phosphate buffered saline (PBS) at pH 7.4, was observed to be 350 nm.[16] Polyurethane nanoparticles observed by AFM were spherical with diameter around 209 nm for nanoparticles prepared without PEG.[17]

Published by Woodhead Publishing Limited, 2013

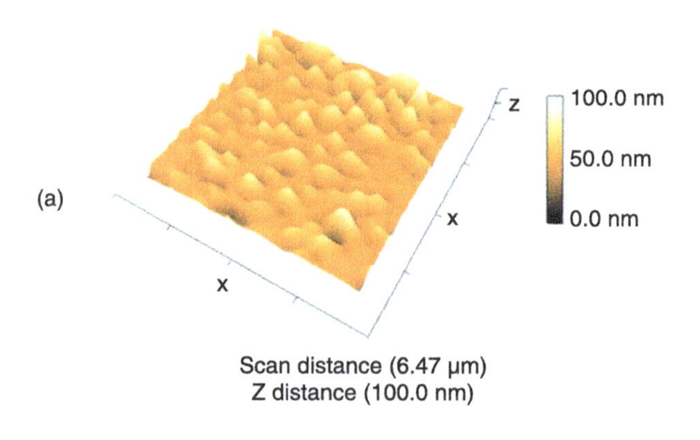

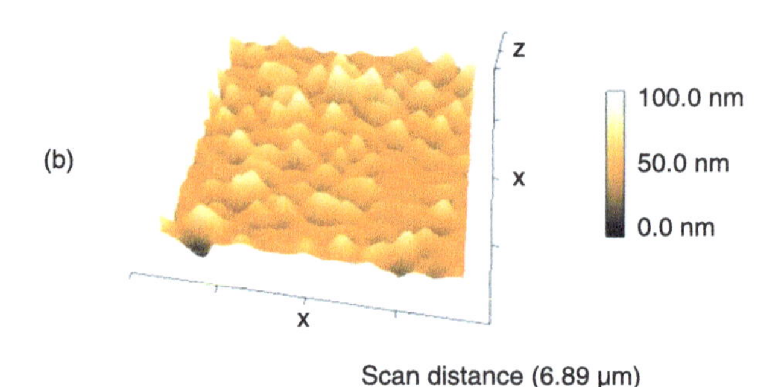

Figure 3.2 Atomic force microscope image of (a) PPA 3 alone and (b) PPA 3-DNA complexes. Average size in each case is 100 nm

Source: adapted from Nimesh et al. 2007[15]

From AFM imaging two modes of nanoparticles were observed in the formulation prepared with PEG, 218 and 127 nm.

3.2.1.2 Dynamic light scattering

DLS measures the temporal fluctuations of the light scattered due to the Brownian motion of the particles, when a solution

Published by Woodhead Publishing Limited, 2013

containing the particles is placed in the path of a monochromatic beam of light. It is also known as photon correlation spectroscopy or quasi-elastic light scattering. This technique provides particle size information in terms of hydrodynamic diameter along with the polydispersity index of the sample. DLS is a sensitive, non-intrusive, and powerful analytical tool, routinely employed for characterization of macromolecules and colloids in solution. This technique has also been used to measure the size and size distribution profile of nanoparticles. In a study published earlier, size of polymer/DNA complexes was determined by DLS fitted with an argon ion laser operated at 633 nm as the light source using a digital correlator.[16] Measurements were carried out at an angle perpendicular to the incident light and the data were collected over a period of 3 min. The mean particle size of the complexes of guanidinated-PEI/pDNA prepared at N/P ratio of 30 in PBS at pH 7.4 was observed to be 355 nm. In another study, PEI–PEG nanoparticles were characterized by DLS and TEM and found to be in the range of 18–75 nm (hydrodynamic radii) with almost uniform population[18] (Figure 3.3).

Usage of two different methods, namely, AFM or TEM and DLS, results in probable discrepancy in the nanoparticle size determination. To perform DLS, nanoparticles are suspended in water or buffer, making them completely hydrated, whereas for AFM or TEM samples are dried on a glass slide or copper grid surface. AFM and TEM have the limitation of visualizing a small number of nanoparticles, while the DLS technique provides an average picture of the sample by determining the size of millions of particles. Therefore, the use of two different but complementary methods provides an overall evaluation of both size and morphology of nanoparticles. However, size determination under physiological conditions is an important

Published by Woodhead Publishing Limited, 2013

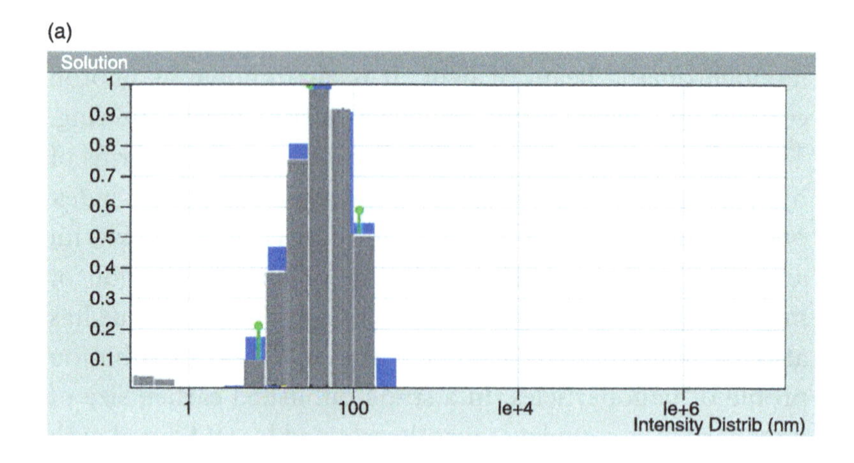

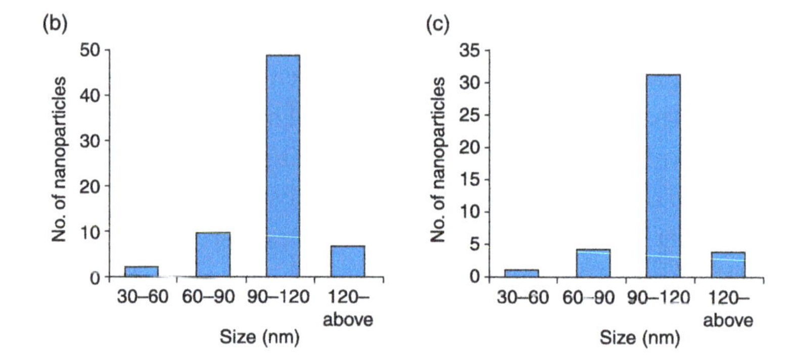

Figure 3.3 (a) Representative overlay of dynamic light scattering spectrum of PEI-PEG_{8000} 10% (ionically cross-linked) nanoparticles in double distilled water (blue color), and 10% FCS (gray color). Size distribution obtained by TEM images of ionic nanoparticles: (b) PEI-PEG_{6000} 10% ionic alone and (c) PEI-PEG_{6000} 10% ionic DNA loaded. Average size in each case is 100 nm

Source: adapted from Nimesh et al. 2006[18]

and challenging area. The properties of nanoparticles are highly influenced by the surrounding environment; for instance, the size distribution at physiological conditions may differ from that in water or in the dry state. In this

Published by Woodhead Publishing Limited, 2013

regard, DLS seems to be the more suitable method, as it provides measurements in physiological buffers or biological fluids such as blood plasma.

3.2.1.3 Laser diffraction

In addition to microscopy and DLS, laser diffraction, which works on the Mie and Fraunhofer scattering theory, has also been employed for particle size measurement. These laser diffraction instruments are quite capable of measuring particle size in the range of a few nanometers to several microns. Moreover, the unique advantage of this technology is not only the wide size range but also its capability to measure dry powders and aerosols, which is not possible by DLS. In a typical experiment, a dispersed sample is passed through the measurement chamber, where a laser beam illuminates the particles. The intensity of light scattered by the particles within the sample is detected over a wide range of angles. In contrast to DLS, the laser diffraction technique can also measure the volume distribution directly, whereas the DLS measures an intensity-based mean particle diameter.

3.2.2 Surface morphology

Surface morphology and shape of nanoparticles play a vital role in interaction with the cells and resulting uptake. TEM, AFM and SEM have been used for determination of the surface morphology of nanoparticles. SEM employs a high-energy beam of electrons to scan the sample surfaces. On this instrument samples are analyzed after drying and coating with a thin layer of gold or platinum. SEM provides a direct picture of the surface of nanoparticles. The surface morphology of PLGA nanoparticles modified with L-ascorbic acid was analyzed by SEM.[19] The synthesized PLGA nanoparticles were

Published by Woodhead Publishing Limited, 2013

found to have regular and smooth spherical shape, with no significant difference between particles prepared with and without ascorbic acid. In another study, Wu et al. investigated (PEG-PEI)/siRNA nanoparticles at N/P ratio of 15 by SEM for morphology.[20] The nanoparticles were observed to be spherical, uniform in size and well dispersed, with average size of 240 nm. In one of our studies, surface morphology of nanoparticles prepared by cross-linking AADG with MBA was determined with TEM.[12] The nanoparticles appeared spherical with smooth surface and uniform distribution with average size of 85 nm (Figure 3.1). AFM has also been employed for investigation of surface morphology of nanoparticles in several studies. We employed AFM to observe nanoparticles prepared from PEI acylated with propionic anhydride with or without complexation with siRNA (Figure 3.4).[21] AFM investigation of these nanoparticles showed spherical and compact complexes

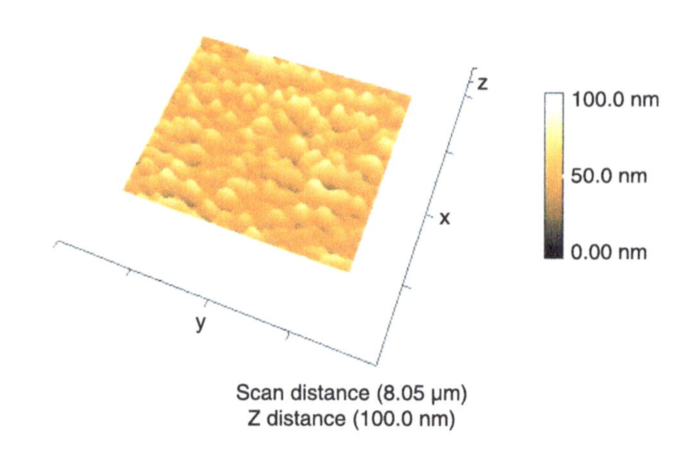

Figure 3.4 AFM image of nanoparticles prepared from PEI acylated with propionic anhydride in double distilled water. The average particle size is 100 nm

Source: adapted from Nimesh and Chandra 2009[21]

Published by Woodhead Publishing Limited, 2013

with an average size of 100 nm. The three-dimensional image revealed a homogeneous population with a clear absence of aggregates even after 2 h. The nanoparticles maintained their morphology and shape even after binding to siRNA.

3.2.3 Surface charge

The surface charge plays an important role in determining the interaction of nanoparticles with the cell surface. Zeta potential is used to describe the surface charge of nanoparticles, which can be defined as a measure of the magnitude of repulsion or attraction between particles. A particle suspended in a solution containing ions is surrounded by an electrical double layer of ions and counterions. The potential that exists at the hydrodynamic boundary of the particle is known as the zeta potential. It is determined using laser Doppler velocimetry, whereby the velocity of particles is measured in samples exposed to electrophoresis. The stability of the colloidal system can be predicted by the magnitude of the zeta potential values. Particles in suspension with a high negative or positive zeta potential will tend to repel each other, thereby resulting in a stable colloid. However, on the contrary, particles with low zeta potential values lack sufficient repulsive force, which results in particle agglomeration. The charge of chitosan/DNA nanoparticles depends on the concentration of DNA and chitosan as well as the pH and salt content of the suspension medium. We systematically investigated the influence of salt and pH on the surface charge of chitosan/DNA nanoparticles.[22] Zeta potential measurements indicated that particle surface charge was reduced by suspending nanoparticles in PBS at pH 6.5 (+11.4 mV) as compared with water at pH 6.1 (+41.4 mV), and even became negative at pH 7.4 (–4.9 mV). This dependence of charge on pH was also observed by Mao et al., who found electrostatically neutral particles in the pH range of 7.0–7.4 using a N/P ratio of 6 while the zeta potential became −20 mV at pH 8–8.5.[23]

Nanomaterial properties such as solubility and cell surface interactions can be significantly influenced by the presence of functional groups on the surface. Incubation of nanomaterials with cells in media results in adsorption of serum proteins on their surface that induce the entry of nanoparticles by receptor-mediated endocytosis.[24] However, synthesis of nanoparticles that do not interact with cell membranes or other biological macromolecules is also desirable. For instance, for *in vivo* applications, non-specific adsorption of proteins on the nanoparticle surface can occur, which may lead to particle agglomeration and clearance by the RES, hindering the particle's ability to deliver drugs/genes to the target site. To avoid such issues, nanoparticles can be coated with neutral polymers such as PEG, which is well known to resist protein adsorption.

3.2.4 Stability

To be beneficial as therapeutic agents, nanoparticles must satisfy two important criteria: they must be stable over the temperature range and the time period required as per the need for therapeutic applications. The surface of nanoparticles needs to be modified in order to improve upon their dispersion stability in liquid media, by polymeric surfactants or other modifiers, so as to generate an effective repulsive force between the nanoparticles. Adsorption of polymeric dispersant is the easiest way to modify the nanoparticle surface. Various carboxylic acid-based anionic surfactants, such as polyacrylic acid and polyacrylic acid sodium salts, and the cationic surfactant PEI are widely accepted.[25–27] However, the high positive charge arising due to amine groups, as in the case of nanoparticles prepared from polycationic polymers, is sufficient to maintain the dispersion stability. Stable shelf life over a period of time is one of the important prerequisites for production of biologically active molecules. It is well established that very small particles tend to aggregate among themselves to reduce the surface area, and hence to reduce the free surface energy. One of the methods to check the stability of the nanoparticles is to examine the particle size at various time points. In one of our recent publications, we have prepared chitosan nanoparticles by cross-linking with glutaraldehyde in reverse micro-emulsion and the stability was examined in terms of the particle size.[28] The lyophilized nanoparticle powder was stored at 4°C for 1–30 days and the size of the nanoparticles monitored by DLS. A clear suspension was obtained by sonication when the lyophilized powder of nanoparticles was dispersed in water. The conclusion of the study suggested that the nanoparticles were quite stable at 4°C, since the size of the nanoparticles remained constant even after 30 days of storage.

Plasmid DNA formulations are prepared for transfections, and these can be stabilized by drying at room temperature, though they are unstable when stored in solution.[29] In a few studies dealing with the applications of PEI-based formulations with siRNA, oligonucleotides and ribozymes, lyophilization has been reported.[30,31] It was observed that these systems retained activity when freeze-dried with glucose. It has been shown that nucleotide complexes lose transfection activity because of interactions between the complexes during the freezing step that leads to aggregation.[32] However, steric hindrance by macromolecules like sucrose could lead to decreased mobility of the particles and so decrease the interactions as well. Andersen et al.

Published by Woodhead Publishing Limited, 2013

also observed a requirement for a lyoprotectant in their freeze-dried chitosan system to prevent aggregation and loss of silencing activity.[33]

3.2.5 Sterility

Maintenance of a sterile atmosphere for the nanoparticles is a prerequisite for both *in vitro* and *in vivo* application in the field of drug and gene delivery. For clinical use, the pharmacopoeial requirements of sterility have to be met by parenteral drug delivery systems. However, the chemical or physical properties of the polymer matrix usually limit most conventional methods for obtaining acceptable sterile products.[34] A major challenge in the case of polymer nanoparticles used as drug carriers is the use of a sterilization technique which can keep intact the supramolecular and molecular structure of the colloids. With chemical sterilization by gases such as ethylene oxide, toxicological problems may be encountered due to toxic residues. Numerous studies have shown the effects of γ-irradiation on the stability and safety of colloidal carriers based on polyesters, principally microparticles and nanoparticles.[35,36] Therefore, the selection of a suitable sterilization method for this type of formulation is crucial to ensure their physical and chemical integrity, their performance, and biosafety *in vivo*.

Sterile filtration based on physical removal of micro-organisms through 0.22-μm membrane filters may be considered the appropriate method for chemically or thermally sensitive materials, since it has no adverse effect on the polymer or the drug.[37–39] Moreover, it has no adverse effect on either the drug release properties and the stability of a formulation or the chemical stability of ingredients. However, this sterilization method can only be used for nanoparticles with a mean size significantly below

Published by Woodhead Publishing Limited, 2013

membrane cut-off and with a narrow size distribution to avoid membrane clogging. Moreover, this technique is not suitable for larger nanoparticles (exceeding 200 nm), when the drug is adsorbed at the nanoparticle surface, or when the colloidal suspensions are too viscous.[40,41] In one of our studies, we sterilized chitosan solutions by a sterile filtration technique before complexation with DNA followed by incubation and transfection of complexes in sterile conditions.[22] The complexes showed efficient transfection without any signs of contamination by either bacteria or mycoplasmas.

Autoclaving for heat-induced sterilization is a highly effective technique, which involves high temperatures and may influence decomposition or degradation of the active ingredient as well as the nanoparticle material, i.e. the polymer.[35,42] A significant increase in particle size was reported after autoclaving unloaded polybutylcyanoacrylate nanoparticle suspensions, and the nanoparticle powders were characterized by impaired re-suspension characteristics. These effects were attributed to swelling of the polymeric membrane. Sterilization by autoclaving induces degradation of polyesters by hydrolysis, and these polymers are also heat-sensitive, due to their thermoplastic nature. On the other hand, solid lipid nanoparticles can be sterilized by autoclaving, maintaining an almost spherical shape, without any significant increase in size or distribution.[43,44] The Food and Drug Administration requires that sterile pharmaceutical products be free of viable micro-organisms. Sterility testing of pharmaceutical products provides added assurance that the product is sterile. Sterility testing is typically done by inoculating the drug product into microbial growth media for Gram-positive and Gram-negative bacteria, spore-forming bacteria, yeasts and fungi, followed by visual inspection for growth during incubation for a specified time period. A lack of visual growth indicates that the drug product samples tested were sterile.

3.3 References

1. Decuzzi, P., Pasqualini, R., Arap, W. and Ferrari, M. (2009) Intravascular delivery of particulate systems: does geometry really matter? *Pharm Res*, **26**, 235–43.
2. Patil, Y.B., Toti, U.S., Khdair, A., Ma, L. and Panyam, J. (2009) Single-step surface functionalization of polymeric nanoparticles for targeted drug delivery. *Biomaterials*, **30**, 859–66.
3. Ogris, M. and Wagner, E. (2002) Targeting tumors with non-viral gene delivery systems. *Drug Discov Today*, **7**, 479–85.
4. Dauty, E., Remy, J.S., Blessing, T. and Behr, J.P. (2001) Dimerizable cationic detergents with a low cmc condense plasmid DNA into nanometric particles and transfect cells in culture. *J Am Chem Soc*, **123**, 9227–34.
5. Lee, H., Williams, S.K., Allison, S.D. and Anchordoquy, T.J. (2001) Analysis of self-assembled cationic lipid-DNA gene carrier complexes using flow field-flow fractionation and light scattering. *Anal Chem*, **73**, 837–43.
6. Sakurai, F., Inoue, R., Nishino, Y., Okuda, A., Matsumoto, O., et al. (2000) Effect of DNA/liposome mixing ratio on the physicochemical characteristics, cellular uptake and intracellular trafficking of plasmid DNA/cationic liposome complexes and subsequent gene expression. *J Control Release*, **66**, 255–69.

Published by Woodhead Publishing Limited, 2013

7. Desai, M.P., Labhasetwar, V., Walter, E., Levy, R.J. and Amidon, G.L. (1997) The mechanism of uptake of biodegradable microparticles in Caco-2 cells is size dependent. *Pharm Res*, **14**, 1568–73.
8. Zauner, W., Farrow, N.A. and Haines, A.M. (2001) In vitro uptake of polystyrene microspheres: effect of particle size, cell line and cell density. *J Control Release*, **71**, 39–51.
9. Prabha, S., Zhou, W.Z., Panyam, J. and Labhasetwar, V. (2002) Size-dependency of nanoparticle-mediated gene transfection: studies with fractionated nanoparticles. *Int J Pharm*, **244**, 105–15.
10. Chithrani, B.D., Ghazani, A.A. and Chan, W.C. (2006) Determining the size and shape dependence of gold nanoparticle uptake into mammalian cells. *Nano Lett*, **6**, 662–8.
11. Chithrani, B.D. and Chan, W.C. (2007) Elucidating the mechanism of cellular uptake and removal of protein-coated gold nanoparticles of different sizes and shapes. *Nano Lett*, 7, 1542–50.
12. Nimesh, S., Manchanda, R., Kumar, R., Saxena, A., Chaudhary, P., et al. (2006) Preparation, characterization and in vitro drug release studies of novel polymeric nanoparticles. *Int J Pharm*, **323**, 146–52.
13. Patnaik, S., Sharma, A.K., Garg, B.S., Gandhi, R.P. and Gupta, K.C. (2007) Photoregulation of drug release in azo-dextran nanogels. *Int J Pharm*, **342**, 184–93.
14. Montasser, I., Fessi, H. and Coleman, A.W. (2002) Atomic force microscopy imaging of novel type of polymeric colloidal nanostructures. *Eur J Pharm Biopharm*, **54**, 281–4.
15. Nimesh, S., Aggarwal, A., Kumar, P., Singh, Y., Gupta, K.C., et al. (2007) Influence of acyl chain length on

Published by Woodhead Publishing Limited, 2013

transfection mediated by acylated PEI nanoparticles. *Int J Pharm*, **337**, 265–74.
16. Nimesh, S. and Chandra, R. (2008) Guanidinium-grafted polyethylenimine: An efficient transfecting agent for mammalian cells. *Eur J Pharm Biopharm*, **68**, 647–55.
17. Zanetti-Ramos, B.G., Fritzen-Garcia, M.B., de Oliveira, C.S., Pasa, A.A., Soldi, V., et al. (2009) Dynamic light scattering and atomic force microscopy techniques for size determination of polyurethane nanoparticles. *Mater Sci Eng C*, **29**, 638–40.
18. Nimesh, S., Goyal, A., Pawar, V., Jayaraman, S., Kumar, P., et al. (2006) Polyethylenimine nanoparticles as efficient transfecting agents for mammalian cells. *J Control Release*, **110**, 457–68.
19. Martins, D., Frungillo, L., Anazzetti, M.C., Melo, P.S. and Duran, N. (2010) Antitumoral activity of L-ascorbic acid-poly-D,L-(lactide-co-glycolide) nanoparticles containing violacein. *Int J Nanomedicine*, **5**, 77–85.
20. Wu, Y., Wang, W., Chen, Y., Huang, K., Shuai, X., et al. (2010) The investigation of polymer-siRNA nanoparticle for gene therapy of gastric cancer in vitro. *Int J Nanomedicine*, **5**, 129–36.
21. Nimesh, S. and Chandra, R. (2009) Polyethylenimine nanoparticles as an efficient in vitro siRNA delivery system. *Eur J Pharm Biopharm*, **73**, 43–9.
22. Nimesh, S., Thibault, M.M., Lavertu, M. and Buschmann, M.D. (2010) Enhanced gene delivery mediated by low molecular weight chitosan/DNA complexes: effect of pH and serum. *Mol Biotechnol*, **46**, 182–96.
23. Mao, H.Q., Roy, K., Troung-Le, V.L., Janes, K.A., Lin, K.Y., et al. (2001) Chitosan-DNA nanoparticles as gene carriers: synthesis, characterization and

Published by Woodhead Publishing Limited, 2013

transfection efficiency. *J Control Release*, **70**, 399–421.

24. Khan, J.A., Pillai, B., Das, T.K., Singh, Y. and Maiti, S. (2007) Molecular effects of uptake of gold nanoparticles in HeLa cells. *Chembiochem*, **8**, 1237–40.
25. Prabhakaran, K., Kumbhar, C.S., Raghunath, S., Gokhale, N.M. and Sharma, S.C. (2008) Effect of concentration of ammonium poly(acrylate) dispersant and MgO on coagulation characteristics of aqueous alumina direct coagulation casting slurries. *J Am Ceram Soc*, **91**, 1933–8.
26. Sato, K., Kondo, S., Tsukada, M., Ishigaki, T. and Kamiya, H. (2007) Influence of solid fraction on the optimum molecular weight of polymer dispersants in aqueous TiO2 nanoparticle suspensions. *J Am Ceram Soc*, **90**, 3401–6.
27. Laarz, E., Meurk, A., Yanez, J.A. and Bergström, L. (2001) Silicon nitride colloidal probe measurements: interparticle forces and the role of surface-segment interactions in poly(acrylic acid) adsorption from aqueous solution. *J Am Ceram Soc*, **84**, 1675–82.
28. Manchanda, R. and Nimesh, S. (2010) Controlled size chitosan nanoparticles as an efficient, biocompatible oligonucleotides delivery system. *J Appl Polym Sci*, **118**, 2071–7.
29. Romoren, K., Aaberge, A., Smistad, G., Thu, B.J. and Evensen, O. (2004) Long-term stability of chitosan-based polyplexes. *Pharm Res*, **21**, 2340–6.
30. Werth, S., Urban-Klein, B., Dai, L., Hobel, S., Grzelinski, M., et al. (2006) A low molecular weight fraction of polyethylenimine (PEI) displays increased transfection efficiency of DNA and siRNA in fresh or lyophilized complexes. *J Control Release*, **112**, 257–70.

Published by Woodhead Publishing Limited, 2013

31. Brus, C., Kleemann, E., Aigner, A., Czubayko, F. and Kissel, T. (2004) Stabilization of oligonucleotide-polyethylenimine complexes by freeze-drying: physicochemical and biological characterization. *J Control Release*, **95**, 119–31.
32. Allison, S.D., Molina, M.C. and Anchordoquy, T.J. (2000) Stabilization of lipid/DNA complexes during the freezing step of the lyophilization process: the particle isolation hypothesis. *Biochim Biophys Acta*, **1468**, 127–38.
33. Andersen, M.O., Howard, K.A., Paludan, S.R., Besenbacher, F. and Kjems, J. (2008) Delivery of siRNA from lyophilized polymeric surfaces. *Biomaterials*, **29**, 506–12.
34. Athanasiou, K.A., Niederauer, G.G. and Agrawal, C.M. (1996) Sterilization, toxicity, biocompatibility and clinical applications of polylactic acid/polyglycolic acid copolymers. *Biomaterials*, **17**, 93–102.
35. Memisoglu-Bilensoy, E. and Hincal, A.A. (2006) Sterile, injectable cyclodextrin nanoparticles: effects of gamma irradiation and autoclaving. *Int J Pharm*, **311**, 203–8.
36. Volland, C., Wolff, M. and Kissel, T. (1994) The influence of terminal gamma-sterilization on captopril containing poly(d,l-lactide-co-glycolide) microspheres. *J Control Release*, **31**, 293–305.
37. Goldbach, P., Brochart, H., Wehrlé, P. and Stamm, A. (1995) Sterile filtration of liposomes: Retention of encapsulated carboxyfluorescein. *Int J Pharm*, **117**, 225–30.
38. Konan, Y.N., Gurny, R. and Allémann, E. (2002) Preparation and characterization of sterile and freeze-dried sub-200 nm nanoparticles. *Int J Pharm*, **233**, 239–52.

Published by Woodhead Publishing Limited, 2013

39. Maksimenko, O., Pavlov, E., Toushov, E., Molin, A., Stukalov, Y., et al. (2008) Radiation sterilisation of doxorubicin bound to poly(butyl cyanoacrylate) nanoparticles. *Int J Pharm*, **356**, 325–32.
40. Brigger, I., Armand-Lefevre, L., Chaminade, P., Besnard, M., Rigaldie, Y., et al. (2003) The stenlying effect of high hydrostatic pressure on thermally and hydrolytically labile nanosized carriers. *Pharm Res*, **20**, 674–83.
41. Tsukada, Y., Hara, K., Bando, Y., Huang, C.C., Kousaka, Y., et al. (2009) Particle size control of poly(dl-lactide-co-glycolide) nanospheres for sterile applications. *Int J Pharm*, **370**, 196–201.
42. Wörle, G., Siekmann, B., Koch, M.H.J. and Bunjes, H. (2006) Transformation of vesicular into cubic nanoparticles by autoclaving of aqueous monoolein/poloxamer dispersions. *Eur J Pharm Sci*, **27**, 44–53.
43. Cavalli, R., Caputo, O., Carlotti, M.E., Trotta, M., Scarnecchia, C., et al. (1997) Sterilization and freeze-drying of drug-free and drug-loaded solid lipid nanoparticles. *Int J Pharm*, **148**, 47–54.
44. Sanna, V., Kirschvink, N., Gustin, P., Gavini, E., Roland, I., et al. (2004) Preparation and in vivo toxicity study of solid lipid microparticles as carrier for pulmonary administration. *AAPS PharmSciTech*, **5**, e27.

Published by Woodhead Publishing Limited, 2013

4

Characterization of nanoparticles: *in vitro* and *in vivo*

DOI: 10.1533/9781908818645.65

Abstract: Efficacy of gene and drug delivery is dictated by the interaction of nanoparticles with the cell membranes. Since the cell membranes are anionic in nature, cationic nanoparticles enhance the interaction with the cell membranes. Binding and uptake of nanoparticles have been characterized by sophisticated microscopy and flow cytometry techniques. Binding of nanoparticles depends on the particles' charge, size, shape and the cell type under investigation. Cytotoxicity has been evaluated by employing a multitude of colorimetric assays exploiting either the physical state or the metabolic state of the cells. Further, *in vivo* characterization of interaction of nanoparticles is highly desirable in deciphering the interaction of nanoparticles with various serum proteins. This chapter explores various factors dictating the characterization of nanoparticles in different milieu and technique employed.

Key words: binding, uptake, confocal microscopy, flow cytometry, cytotoxicity, alarmar blue, MTT assay, western blot.

Published by Woodhead Publishing Limited, 2013

4.1 Introduction

Interaction with the cell membrane is a prerequisite for nanoparticles, in order to be efficient as a gene and drug carrier to the target. Therefore, an insight into biophysical interactions of nanoparticles with the cell is necessary if they are to be designed and developed as a potent carrier system for therapeutic and pharmaceutical research applications. Electrostatic force is potentially the major force of interaction between the nanoparticles and the cellular membrane. Several proteins are either completely or partially embedded in the plasma membrane, and these are responsible for the negative charge on the membrane, interacting efficiently with the positively charged nanoparticles. This is the reason why cationic polymers are preferred, since a positive charge on the nanoparticles' surface is highly advantageous. But cytotoxicity due to access of the positive charge limits the usage of cationic polymers. Thus, various modifications and characterization of nanoparticles are required in order to develop these as vectors with desirable qualities, keeping in mind that the interaction of nanoparticles, followed by their delivery activity, is a crucial, multi-step process.

4.2 *In vitro* characterization of nanoparticles

4.2.1 *Binding and uptake*

Binding of nanoparticles to the cellular membrane is the first and one of the most important steps, followed by uptake. Cell surface proteins and sugar moieties on the membrane are usually negatively charged, and cationic nanoparticles

Published by Woodhead Publishing Limited, 2013

with size less than a micron interact with these negatively charged species. These kinds of interactions are usually non-specific, but they stimulate the onset of the endocytic pathway for engulfment by endosomal vesicles into the cell, a process known as endocytosis. Various extra-endosomal factors and the cytoskeleton strictly define the life cycle of endosomes and regulate the multi-step process of endocytosis. The various steps involved in the process begin with the engulfment of material in membrane invaginations which are pinched off to form membrane-bound vesicles, also known as endosomes (or phagosomes in the case of phagocytosis). Cells are equipped with heterogeneous populations of endosomes that possess distinct endocytic machinery and originate at various sites of the cell membrane. To continue with the endocytic process, sorting of materials towards different destinations is required, and this is done by endosomes delivering the material to various specialized vesicular structures. Finally, the material is delivered to various intracellular compartments, recycled to the extracellular milieu or delivered across cells (a process known as "transcytosis" in polarized cells). Generally, endocytosis is categorized into two broad classes: phagocytosis (the uptake of large particles) and pinocytosis (the uptake of fluids and solutes).

Advanced and sophisticated techniques such as flow cytometry and confocal microscopy have been employed in various research studies to investigate the interaction of nanoparticles with the cells. In one of our studies, flow cytometry was employed to investigate the uptake of chitosan/DNA nanoparticles by HEK 293 cells.[1] Cellular uptake of nanoparticles was quantified by flow cytometry; however, it does not generally discriminate between membrane-bound and internalized fluorescent moieties, so additional procedures were included to minimize any

Published by Woodhead Publishing Limited, 2013

contribution of surface-bound particles in measuring uptake. For this reason, cells were treated with trypsin, followed by several washes, before fluorescence assisted cell sorting (FACS) analysis, which proved to be a quick and efficient method to remove surface-bound complexes and assess cellular uptake of fluorescent chitosan and complexes. Uptake of complexes by HEK 293 cells was pH and serum-dependent, with maximum uptake occurring in medium at pH 6.5 supplemented with 10% fetal bovine serum (FBS). The uptake of nanoparticles as well as of chitosan alone was higher in the presence of serum at all the three pH values investigated.[1] The interaction between chitosan and the plasma membrane is due to non-specific electrostatic forces of attraction; no receptor specific to chitosan has been identified in the cell membrane.[2] Further, to fully substantiate the uptake results from flow cytometry analysis, we assessed the cellular internalization of nanoparticles with confocal microscopy.[1] Though this technique is more qualitative than quantitative, it permits direct observation of the localization of the fluorescent nanoparticles in the cells, providing evidence of cellular internalization. During cellular tracking of the fluorescent chitosan/DNA nanoparticles, fluorescence appeared to be distributed throughout the cells at pH 6.5 with serum (Figure 4.1). However, large aggregates of complexes were found near the cell membranes with only small amounts being internalized when cells were incubated with the complexes at pH 7.1 and 7.4 with serum.[1]

We observed strong correlation between cell binding (Figure 4.2(a)) and cell uptake (Figure 4.2(b)), with a notable increase in uptake for the two chitosans with high degree of deacetylation (DDA) (92%) compared with lower-DDA chitosans.[3] The correlation between binding and uptake

Published by Woodhead Publishing Limited, 2013

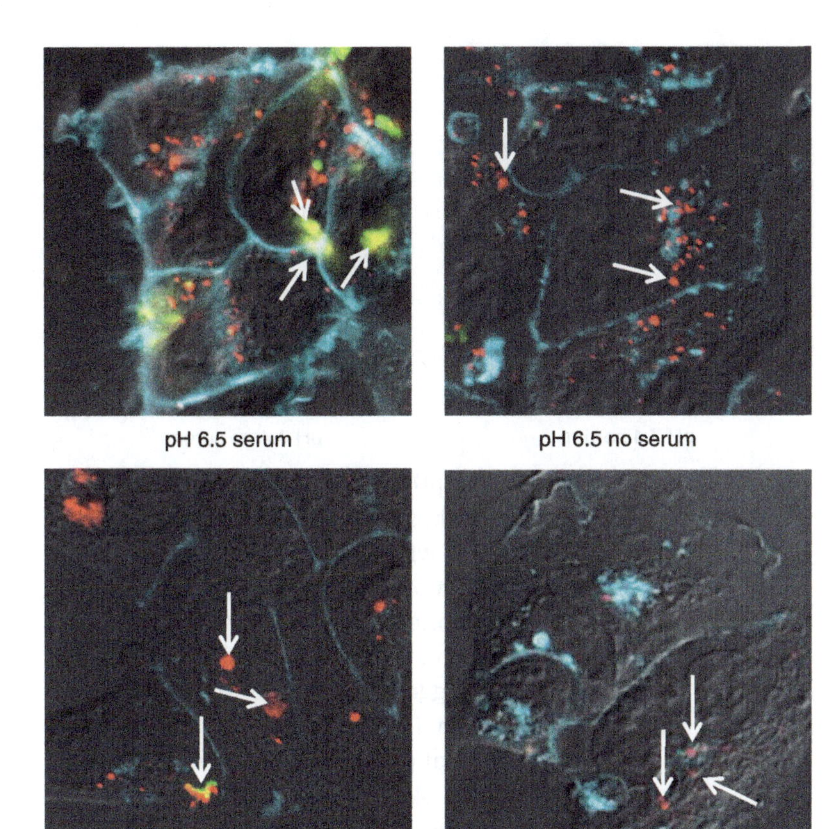

Figure 4.1 Confocal microscope images of HEK 293 cells transfected at different pH values. Cells transfected with chitosan/DNA complexes, where chitosan is labeled with rhodamine and DNA with fluorescein, visualized under confocal microscope 24 h post-transfection. Blue color: staining of cell membrane; red color: rhodamine-labeled chitosan only; green color: fluorescein-labeled DNA only; yellow: colocalization showing chitosan/DNA complex. At pH 6.5 with serum, a large number of complexes are visible inside the cell (white arrows) whereas at pH 6.5 without serum a large amount of free chitosan can be seen inside the cell (white arrows).

(continued)

Published by Woodhead Publishing Limited, 2013

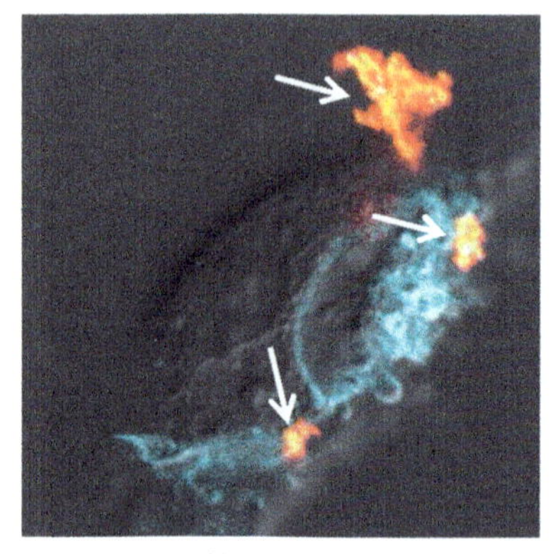

pH 7.4 serum

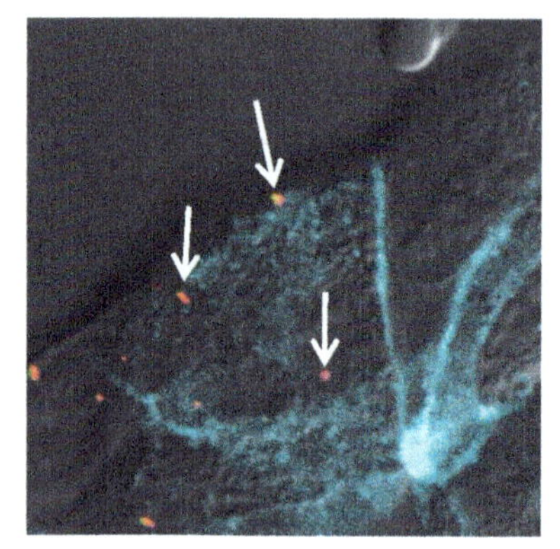

pH 7.4 no serum

Figure 4.1 (*continued*) At pH 7.1 with serum, small amounts of chitosan and complexes were visible (white arrows), whereas at pH 7.1 without serum only a small amount of free chitosan was seen inside cells (white arrows) without any internalization of complexes. At pH 7.4 with serum, large aggregates were observed outside the cells external to the cell membrane (white arrows), while at pH 7.4 without serum a small number of complexes were observed on cell membranes (white arrows)

Source: adapted from Nimesh et al. 2010[1]

underscores the necessity of establishing cell contact with polyplexes. The number of cells positive for uptake (Figure 4.2(c)) peaked at around 8 hours, when nearly 100% of the cells contained internalized polyplexes. In contrast, the total amount being taken per cell increased continuously with time till the medium was changed, indicating that no saturation of internalization occurs at any time point. Several parameters that define the success of the process of internalization of nanoparticles are reported and discussed below.

Published by Woodhead Publishing Limited, 2013

(a)

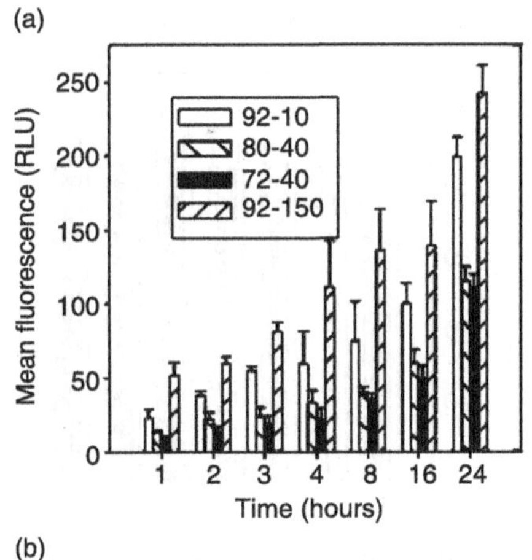

(b)

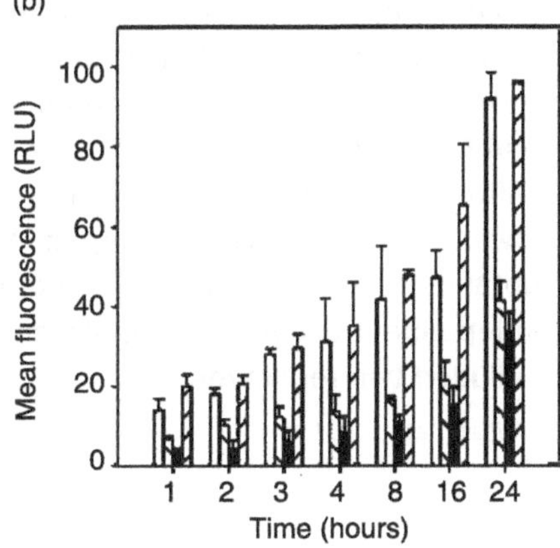

Figure 4.2 Kinetics of cellular binding and uptake of polyplexes prepared with chitosans of different DDA and MW. HEK293 cells were incubated with fluorescent chitosan polyplexes for the indicated periods of time and analyzed by flow cytometry. Flow cytometry quantitative analysis of mean fluorescence per cell for polyplex (a) binding and (b) uptake and (c) of % cells with internalized polyplexes was performed following trypsinization and extensive washes, except for (a) cell binding, where cells were detached by enzyme-free cell dissociation buffer and analyzed directly. (b) Mean uptake levels per cell and (continued)

Published by Woodhead Publishing Limited, 2013

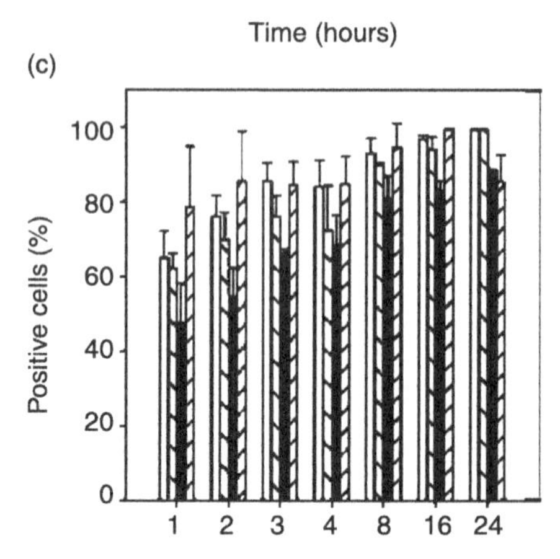

Figure 4.2 (*continued*) (c) % positive cells were obtained from the same set of flow cytometry data. Graphs show that binding and uptake are time and DDA-dependent, with both 92% DDA chitosans binding more effectively than the lower-DDA chitosans, resulting in increased uptake. Results are the average of three ($N=3$) independent experiments ± SD; each experiment included two replicates

Source: adapted from Thibault et al. 2010[3]

4.2.1.1 Particle charge

The surface charge of nanoparticles defines their internalization by cells. At present, the majority of reports suggest that positively charged nanoparticles predominantly internalize through clathrin-mediated endocytosis, with some fraction utilizing macropinocytosis. Examples include cationic nanoparticles of totally different origin—stearylamine-coated PEG-*co*-PLA, PLGA modified with PLL, chitosan, etc.[4,5] However, at times certain strong cationic nanoparticles such as PEI-based polyplexes may utilize multiple pathways including caveolae-mediated endocytosis,[6] while negatively charged nanoparticles, such as DOXIL®,

Published by Woodhead Publishing Limited, 2013

micelles and quantum dots, are more likely to utilize caveolae-mediated endocytosis.[7,8] Slow rate of internalization is associated with negatively charged nanoparticles as compared with their positively charged counterparts because the cell membranes are also generally negatively charged. However, the preference of neutral nanoparticles for specific cellular entry routes is unclear from the available literature.

4.2.1.2 Particle size

The endocytic vesicles vary greatly in size, and thus the size of nanoparticles could play a dominant role in their uptake in different vesicles. This dependency on size led to the idea of the requirement to keep particles small, between 10 and 100 nm, in order to allow the entry of nanoparticles into the endocytic vesicles. It seems that there is no size cut-off limit up to at least 5 μm, though small size may favor rapid entry into cells of some materials through pinocytosis. Also, the largest particles are more likely to enter cells through macropinocytosis.[9] For example, negatively charged polystyrene nanoparticles showed clathrin-mediated endocytosis-based entry for 43-nm particles and caveolae-mediated entry for 25-nm particles.[10] One of the major problems in identifying the effect of the size of nanoparticles is their high polydispersity. This is not true for all nanoparticles; for example, in the case of highly polydispersed chitosan nanoparticles, the chemical composition of the nanomaterial defines the entry pathway, while the effect of particle size is less pronounced.[11]

4.2.1.3 Particle shape

The effect of the shape of particles on phagocytosis is very well described in alveolar rat macrophages in a study by

Published by Woodhead Publishing Limited, 2013

Champion et al.[12] For the study, polystyrene-based particles of more than 20 shapes, including spheres, rectangles, rods, worms, oblate ellipses, elliptical disks and UFO-like, were prepared in the size range primarily from 1 to about 10 μm. However, the size was not a rate-limiting factor in phagocytosis.[13] All particle shapes, independently of their size, were capable of initiating phagocytosis in at least one orientation. However, unexpectedly, the crucial role in phagocytosis was played by the local particle shape at the point of attachment of the particle on the macrophage.[12] For example, if the narrow side of an ellipse is attached to a macrophage, it is internalized in a few minutes, while an ellipse attached by its wider side to the macrophage is not internalized even after several hours. However, spherical particles attached at any point were internalized, independently of their size. Although particle size played a much smaller role in the initiation of phagocytosis, it could, of course, affect its completion, especially when the particle volume exceeded that of a cell. The effect of the geometry on phagocytosis was quantified by measuring the angle between the membrane normal at the point of initial contact and the line defining the particle curvature at this point.[13] The macrophages lose their ability to entrap nanoparticles if this angle exceeds a critical value (45°) and attach to these particles in a spreading kind of process. The authors suggested that the shape of the particle at the point of attachment defines the complexity of actin structures that need to be rearranged to allow engulfment. Above the critical angle the necessary actin structures cannot be created, forcing the macrophages to switch to spreading behavior. Altogether, this eloquent study clearly demonstrated the relationship between the physical properties of the materials to be phagocytosed and cellular transport by the process of phagocytosis.

Published by Woodhead Publishing Limited, 2013

4.2.1.4 Cell type

One of the criteria for cellular uptake and trafficking could be the nature of the cell. But it appears that this relationship, between the nanoparticles' cellular trafficking and cell type, is the least explored. There are few reports that the cell type may be critical in finalizing the entry of nanoparticles and their delivery to the final destination in the cell. It is noteworthy that the differential endocytic pathways in normal and tumor cells may be a gateway for selective targeting of novel nanomaterials into tumors.[7] The relationship between cell origin and availability of various cellular endocytic pathways has not so far been very well considered and emphasized in most of the studies performed on nanoparticle trafficking. It could be possible that, depending on the type of cell, its origin or phenotype, and in some cases even on growing conditions, such as cell density, presence of growth factors, etc., the various known cellular pathways may be differentially presented or even totally absent. Obviously, in-depth understanding of cell biology and its relation to nanomaterials science is critical for developing this area of nanomedicine and drug delivery.

4.2.2 Cytotoxicity

In order to design and develop nanomedicine-based therapeutic agents, the primary concern is to develop an acceptable, efficient and minimally toxic delivery system that ensures maximum patient safety on administration in any form. To target specific cells, nanoparticles are often coated with bioconjugates such as DNA, proteins, and monoclonal antibodies and then used for imaging and drug delivery. Any modifications engineered on the nanoparticles to enhance their interaction with cells must be done with utmost care to

Published by Woodhead Publishing Limited, 2013

ensure that these enhancements do not cause any adverse effects. More significantly, it must be kept in mind that naked or coated nanoparticles will undergo biodegradation once they are in the cellular milieu, and the degraded nanoparticles could induce cellular responses. A plethora of reports have been published reporting *in vitro* cytotoxicity studies of nanoparticles using different cell lines, incubation times, and colorimetric assays. These studies employ a wide range of nanoparticle concentrations and exposure times, making it difficult to determine whether the cytotoxicity observed is physiologically relevant. In addition, different groups have used not only various cell lines but also, at times, different culture conditions, making it difficult to perform direct and conclusive comparisons between the available studies.

To get an understanding about the reaction of an agent in the body, one must conduct cell culture studies with that particular agent as the first step. Cell culture studies must be done as the initiation step because, in comparison to animal studies, the former are less ethically ambiguous; easier to control and reproduce; and less expensive. Cells are sensitive to a number of factors, including any changes in their growing environment such as pH, fluctuations in temperature, nutrient and waste concentrations; and these points must be considered while doing cytotoxicity studies, along with the fact that concentration of the tested agent could also be potentially toxic. Therefore, it is very important to monitor the experimental conditions to ensure that the measured cell death corresponds to the toxicity of the added nanoparticles rather than unstable culture conditions. An appropriate assay for the cytotoxicity test must be chosen, based on the fact that nanoparticles can also adsorb dyes and be redox active. Hence, to draw a valid conclusion from any cytotoxicity study, performing multiple assays is an advantageous and judicious choice.

Published by Woodhead Publishing Limited, 2013

On the onset of any kind of toxicity, the cellular or nuclear morphology is generally changed, so one of the simplest cytotoxicity tests involves visual inspection of the cells with bright field microscopy for changes in cellular or nuclear morphology. This technique was used by us in one of our studies involving PEI nanoparticles, and we observed that PEI polymer induced considerable toxicity and cell morbidity in COS-1 cells as compared with PEI–PEG nanoparticles.[14] Another qualitative method to observe cytotoxicity is a cell confluency study. In another of our experiments, we used cell confluency, the qualitative measure of cell viability based upon cell coverage on the well surface, as judged by microscopy, for an indication of toxicity of chitosan/DNA nanoparticles used for transfection in comparison to Lipofectamine.[1] It was found that, on transfection with chitosan/DNA nanoparticles, the cells were more confluent and retained their normal shape, in contrast to Lipofectamine. Cell membranes become leaky and cellular contents could ooze out when cells are exposed to certain cytotoxic agents, as the membrane could be compromised, making it more permeable. Many commercially available test kits for cytotoxicity have exploited this increased cellular membrane permeability factor; for example, neutral red and trypan blue assays. Neutral red or toluylene red is a weak cationic dye that can cross the plasma membrane by diffusion and becomes accumulated in lysosomes. If the cell membrane is altered, there is a decrease in the uptake of neutral red and also the dye leaks out, allowing live and dead cells to be distinguished. Cytotoxicity can be quantified by taking spectrophotometric measurements of the neutral red uptake under varying exposure conditions. Studies by Flahaut et al. and Monteiro-Riviere et al. determining the cytotoxicity of carbon nanotubes utilized the neutral red assay.[15,16] Trypan blue a diazo dye, can enter only those cells whose membrane

Published by Woodhead Publishing Limited, 2013

is compromised and has become more permeable; therefore, staining the dead cells blue while live cells remain colorless. Light microscopy also finds its use in determining the amount of cell death,[17] and this assay in particular was used by Bottini et al. and Goodman et al. to determine the cytotoxicity of single-walled nanotubes and gold nanoparticles.[18,19]

There are several other colorimetric cytotoxicity assays aimed at detecting the metabolic activity of the cells along with those that differentiate between live and dead cells by detecting damaged cell membranes. One such assay to test the mitochondrial activity of cells uses tetrazolium salts, as mitochondrial dehydrogenase enzymes, present in active mitochondria of live cells, cleave the tetrazolium ring.[20] The MTT assay is one of the most widely employed mitochondrial activity-based tests; it is pale yellow in solution but produces a dark blue formazan product within live cells. In our studies dealing with PEI-derivatized nanoparticles, we have also utilized MTT for determination of cytotoxicity.[14,21–24] Recently, the toxicity of nanoparticles was evaluated by colorimetric Alamar blue assay.[25,26] The blue colored reagent Alamar blue contains resazurin, which is reduced to a pink colored resorufin by the metabolic mitochondrial activity of viable cells and can be quantified colorimetrically and fluorimetrically. The Alamar blue assay was employed by us in our studies to assess potential cytotoxicity of nanoparticles prepared from high DDA (92%) and low MW (10 kDa) chitosan.[1] The chitosan/DNA nanoparticles were found to be only slightly toxic, with more than 85% of cells viable at pH 6.5 and 96% at pH 7.1 even at 48 h post-transfection (Figure 4.3).[1] Since nanoparticles can interact with any cellular components, such as plasma membrane, organelles, or macromolecules, there is a plethora of biological effects that can be triggered during application of nanoparticles, and, since different nanoparticles behave distinctly, it is

Published by Woodhead Publishing Limited, 2013

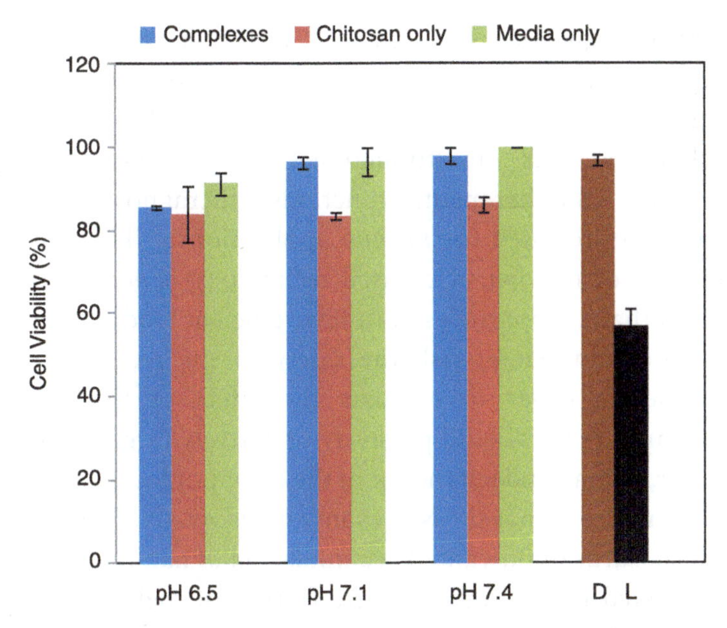

Figure 4.3 Cell viability at different pH values. HEK 293 cells treated with chitosan, chitosan/DNA complexes at different pH, 48 h post-transfection cell viability determined using Alamar blue. Cells at pH 7.4 in complete media taken as 100% viable. Other controls include D: cells incubated with DNA alone, L: cells transfected using Lipofectamine. Values are mean ± SD, $n = 3$

Source: adapted from Nimesh et al. 2010[1]

important that cytotoxicity studies must be conducted for each kind of nanoparticle.

4.3 *In vivo* characterization

Nanoparticles have emerged as a very promising therapeutic agent with a wide range of applications in nanomedicine. Due to their unique characteristics, they have found usage in

Published by Woodhead Publishing Limited, 2013

in vitro diagnostic assays as well as for *in vivo* localized imaging, drug delivery, and therapy.[27–29] There is a drive for insight into *in vivo* biodistribution and possible clearance mechanisms for multifunctional nanoparticles with diagnostic and therapeutic functions ('theragnostics') that have been proposed for *in vivo* applications.[30] The clinical relevance of nanoparticle-based technology is governed by the fact that nanoparticles must have efficient biodistribution that does not put at risk the safety of the recipient. For example, in order to evaluate the effect of therapeutic nanoparticles that could passively or actively target a site of interest, a clear understanding of the extent of localization of nanoparticles at that site is relevant and necessary. Evaluation of the hazards of unwanted accumulation of nanoparticles in off-targeted tissues and characterization of their long-term fate (indefinite residence, metabolism, or excretion) is required for understanding the safety of administered nanoparticles.[31] In view of this important consideration, ever since the synthesis of the first nanoparticle made for biomedical applications, thorough biodistribution studies of nanoparticles have been thoroughly conducted, mainly in relation to polymeric nanoparticles.[32]

Development of nanoparticles is considered to be a boon in the field of nanotherapy, as they can be employed to provide protection or reduce renal clearance for drugs that either have a short life or can be easily degraded, for example, small peptides and nucleic acids, and can, therefore, provide a sustained and prolonged therapeutic effect. However, nanoparticles may also improve the availability of drugs to certain tissues, and thus cause new side effects. The pharmacokinetic profiles of the parent drug and the drug entrapped in the nanoparticles are often different. Hence, monitoring of the pharmacokinetics and biodistribution of nanoparticles is essential to understand and predict their

Published by Woodhead Publishing Limited, 2013

efficacy and side effects. The chemical and physical properties, such as size, charge, and surface chemistry, of the nanoparticles define their pharmacokinetic profile. A pharmacokinetic study involves measuring drug concentrations in all major tissues after drug administration over a period of time until the elimination phase. Monitoring the drug concentration for a long enough period (usually 3 × half-life) is essential to fully describe the behavior of the drug or nanoparticles *in vivo*.

Once the nanoparticles are administered intravenously, a number of serum proteins bind to their surface, which are recognized by the scavenger receptor on the macrophage cell surface and are internalized, leading to a significant removal of nanoparticles from the circulation. This process of binding of the serum proteins to the nanoparticles is also known as opsonization, and the proteins are termed opsonins. One of the major issues in developing a prolonged circulation for a nanoparticle formulation is to minimize the protein interaction and binding to the nanoparticles.

4.3.1 Protein binding to nanoparticles

An enriched environment of proteins, cells and tissues greets the nanoparticles once they are exposed to the bloodstream. Probably the surface of the nanoparticles is initially occupied by the proteins present at high concentrations in plasma and having high association rates. However, over time these proteins may dissociate and be replaced by proteins of lower concentrations, slower exchange rates, and/or higher affinities.[2,6] This whole process is collectively defined as the "Vroman Effect," i.e. the process of competitive adsorption of proteins onto a limited surface based on abundance, affinities, and incubation time.[24] This effect significantly

Published by Woodhead Publishing Limited, 2013

contributes to particle distribution throughout the body. As the particles distribute from the blood to various locations, the differences in protein levels, as well as their affinities for binding, may play a part in determining how the protein corona evolves as the nanoparticle moves from compartment to compartment. The kinetics of protein adsorption can be studied in several ways. Various samples were analyzed in a time-dependent manner with the combined use of SDS-polyacrylamide gel electrophoresis (SDS-PAGE) and western blotting in a study on 50-nm lecithin-coated polystyrene nanospheres. Quantitative and qualitative profiles of serum proteins adsorbed on the surface of the nanoparticles over periods from 5 min to 360 min were revealed in this particular study. SDS-PAGE and western blotting provide vital information about the proteins adsorbed onto the nanoparticles. To get quantitative data from the gel methods is very difficult; these methods are generally used for comparison purposes rather than for precise quantitation. SDS-PAGE can detect anywhere between 1 and 50 ng of a single protein band, depending on the stain used, while western blot limitations are dependent on the antibodies used as well as the conjugated substrate of choice. All of these conditions would have to be optimized for the particular protein of interest. Another study evaluated the kinetics of protein binding to PEG–PHDCA nanoparticles using a system that separates fluorescent dye-labeled protein–SDS complexes electrokinetically in a gel medium, and separates the proteins based on molecular weight.[9] A few established techniques have recently been used to analyze the kinetic properties of protein binding to nanoparticle surfaces in a less potentially disruptive manner than centrifugation. These techniques include size-exclusion chromatography, isothermal titration calorimetry, and surface plasmon resonance.[2] For various poly(N-isopropylacrylamide/N-tert-butylacrylamide)

Published by Woodhead Publishing Limited, 2013

(NIPAM/BAM) copolymer nanoparticles, isothermal titration calorimetry was used to determine the stoichiometry and affinity of human serum albumin (HSA) to the nanoparticle surface.

4.4 Conclusions

The whole concept of designing a drug delivery system is to attain targeted delivery of the drug, not only to a specific cell population but also to a specific intracellular compartment. Nanoparticles can be of immense use; they can either serve as the carrier for drugs or themselves act as therapeutic and imaging agents, thus emphasizing the need to decipher the mechanism of intracellular trafficking of synthetic nanoparticles. Development of nanoparticles directed towards selected intracellular compartments through engagement of specific cellular trafficking machinery is the need of the hour. As various published reports have suggested, the fate and movement of the nanoparticles in the cells depend on a number of factors, such as structure and physico-chemical characteristics of nanoparticles (size, shape, charge, hydrophobicity, etc.), bio-specific interactions between the cell and the biological moieties attached onto nanoparticles, and cell-specific endocytosis pathways. The interaction of the nanoparticles and the cell involves a complex mechanism that also governs the sorting of nanoparticles to different destinations. Also, the interaction between cellular components and synthetic nanoparticles can trigger activation of cellular signaling. Defining the common factors that dictate the trafficking of nanoparticles is a Herculean task due to the diversity of nanoparticles and cells employed in the transport studies. In order to minimize the non-specific uptake of the nanoparticles, avoiding protein binding on the

Published by Woodhead Publishing Limited, 2013

surface and possible engulfment by macrophages, the strategy of coating nanoparticles can be very important for developing a nanoparticle-based therapy that would also provide longer circulation times in the body. On the contrary, to generate nanoparticles that could be targeted or directed to the particular site of localization or a destined pathway, specific bound proteins can be attached to the nanoparticles.[24] Therefore, nanoparticles need to be engineered by specifically binding certain proteins of interest such that they can be used for targeting purposes and escape binding of opsonins to significantly improve their capability to be developed as targeted nanoparticles for therapeutics.

4.5 References

1. Nimesh, S., Thibault, M.M., Lavertu, M. and Buschmann, M.D. (2010) Enhanced gene delivery mediated by low molecular weight chitosan/DNA complexes: effect of pH and serum. *Mol Biotechnol*, **46**, 182–96.
2. Schipper, N.G., Olsson, S., Hoogstraate, J.A., de Boer, A.G., Varum, K.M., et al. (1997) Chitosans as absorption enhancers for poorly absorbable drugs 2: mechanism of absorption enhancement. *Pharm Res*, **14**, 923–9.
3. Thibault, M., Nimesh, S., Lavertu, M. and Buschmann, M.D. (2010) Intracellular trafficking and decondensation kinetics of chitosan-pDNA polyplexes. *Mol Ther*, **18**, 1787–95.
4. Harush-Frenkel, O., Rozentur, E., Benita, S. and Altschuler, Y. (2008) Surface charge of nanoparticles determines their endocytic and transcytotic pathway in polarized MDCK cells. *Biomacromolecules*, **9**, 435–43.

Published by Woodhead Publishing Limited, 2013

5. Harush-Frenkel, O., Debotton, N., Benita, S. and Altschuler, Y. (2007) Targeting of nanoparticles to the clathrin-mediated endocytic pathway. *Biochem Biophys Res Commun*, **353**, 26–32.
6. Rejman, J., Bragonzi, A. and Conese, M. (2005) Role of clathrin- and caveolae-mediated endocytosis in gene transfer mediated by lipo- and polyplexes. *Mol Ther*, **12**, 468–74.
7. Sahay, G., Kim, J.O., Kabanov, A.V. and Bronich, T.K. (2010) The exploitation of differential endocytic pathways in normal and tumor cells in the selective targeting of nanoparticulate chemotherapeutic agents. *Biomaterials*, **31**, 923–33.
8. Zhang, L.W. and Monteiro-Riviere, N.A. (2009) Mechanisms of quantum dot nanoparticle cellular uptake. *Toxicol Sci*, **110**, 138–55.
9. Gratton, S.E., Ropp, P.A., Pohlhaus, P.D., Luft, J.C., Madden, V.J., et al. (2008) The effect of particle design on cellular internalization pathways. *Proc Natl Acad Sci U S A*, **105**, 11613–18.
10. Lai, S.K., Hida, K., Chen, C. and Hanes, J. (2008) Characterization of the intracellular dynamics of a non-degradative pathway accessed by polymer nanoparticles. *J Control Release*, **125**, 107–11.
11. Huang, M., Ma, Z., Khor, E. and Lim, L.Y. (2002) Uptake of FITC-chitosan nanoparticles by A549 cells. *Pharm Res*, **19**, 1488–94.
12. Champion, J.A., Katare, Y.K. and Mitragotri, S. (2007) Making polymeric micro- and nanoparticles of complex shapes. *Proc Natl Acad Sci U S A*, **104**, 11901–4.
13. Champion, J.A. and Mitragotri, S. (2006) Role of target geometry in phagocytosis. *Proc Natl Acad Sci U S A*, **103**, 4930–4.

Published by Woodhead Publishing Limited, 2013

14. Nimesh, S., Goyal, A., Pawar, V., Jayaraman, S., Kumar, P., et al. (2006) Polyethylenimine nanoparticles as efficient transfecting agents for mammalian cells. *J Control Release*, **110**, 457–68.
15. Flahaut, E., Durrieu, M.C., Remy-Zolghadri, M., Bareille, R. and Baquey, C. (2006) Investigation of the cytotoxicity of CCVD carbon nanotubes towards human umbilical vein endothelial cells. *Carbon*, **44**, 1093–9.
16. Monteiro-Riviere, N.A. and Inman, A.O. (2006) Challenges for assessing carbon nanomaterial toxicity to the skin. *Carbon*, **44**, 1070–8.
17. Altman, S.A., Randers, L. and Rao, G. (1993) Comparison of trypan blue dye exclusion and fluorometric assays for mammalian cell viability determinations. *Biotechnol Prog*, **9**, 671–4.
18. Bottini, M., Bruckner, S., Nika, K., Bottini, N., Bellucci, S., et al. (2006) Multi-walled carbon nanotubes induce T lymphocyte apoptosis. *Toxicol Lett*, **160**, 121–6.
19. Goodman, C.M., McCusker, C.D., Yilmaz, T. and Rotello, V.M. (2004) Toxicity of gold nanoparticles functionalized with cationic and anionic side chains. *Bioconjug Chem*, **15**, 897–900.
20. Mosmann, T. (1983) Rapid colorimetric assay for cellular growth and survival: application to proliferation and cytotoxicity assays. *J Immunol Methods*, **65**, 55–63.
21. Nimesh, S., Aggarwal, A., Kumar, P., Singh, Y., Gupta, K.C., et al. (2007) Influence of acyl chain length on transfection mediated by acylated PEI nanoparticles. *Int J Pharm*, **337**, 265–74.
22. Nimesh, S. and Chandra, R. (2009) Polyethylenimine nanoparticles as an efficient in vitro siRNA delivery system. *Eur J Pharm Biopharm*, **73**, 43–9.

Published by Woodhead Publishing Limited, 2013

23. Nimesh, S. and Chandra, R. (2008) Guanidinium-grafted polyethylenimine: an efficient transfecting agent for mammalian cells. *Eur J Pharm Biopharm*, **68**, 647–55.
24. Patnaik, S., Aggarwal, A., Nimesh, S., Goel, A., Ganguli, M., et al. (2006) PEI-alginate nanocomposites as efficient in vitro gene transfection agents. *J Control Release*, **114**, 398–409.
25. Nociari, M.M., Shalev, A., Benias, P. and Russo, C. (1998) A novel one-step, highly sensitive fluorometric assay to evaluate cell-mediated cytotoxicity. *J Immunol Methods*, **213**, 157–67.
26. Nakayama, G.R., Caton, M.C., Nova, M.P. and Parandoosh, Z. (1997) Assessment of the Alamar Blue assay for cellular growth and viability in vitro. *J Immunol Methods*, **204**, 205–8.
27. Agasti, S.S., Rana, S., Park, M.H., Kim, C.K., You, C.C., et al. (2010) Nanoparticles for detection and diagnosis. *Adv Drug Deliv Rev*, **62**, 316–28.
28. Kennedy, L.C., Bickford, L.R., Lewinski, N.A., Coughlin, A.J., Hu, Y., et al. (2011) A new era for cancer treatment: gold-nanoparticle-mediated thermal therapies. *Small*, **7**, 169–83.
29. Veiseh, O., Gunn, J.W. and Zhang, M. (2010) Design and fabrication of magnetic nanoparticles for targeted drug delivery and imaging. *Adv Drug Deliv Rev*, **62**, 284–304.
30. Zolnik, B.S. and Sadrieh, N. (2009) Regulatory perspective on the importance of ADME assessment of nanoscale material containing drugs. *Adv Drug Deliv Rev*, **61**, 422–7.
31. Li, M., Al-Jamal, K.T., Kostarelos, K. and Reineke, J. (2010) Physiologically based pharmacokinetic modeling of nanoparticles. *ACS Nano*, **4**, 6303–17.

Published by Woodhead Publishing Limited, 2013

32. Alexis, F., Pridgen, E., Molnar, L.K. and Farokhzad, O.C. (2008) Factors affecting the clearance and biodistribution of polymeric nanoparticles. *Mol Pharm*, 5, 505–15.

Published by Woodhead Publishing Limited, 2013

5

Theory and limitations to gene therapy

DOI: 10.1533/9781908818645.89

Abstract: Gene or DNA delivery to mammalian cells is a tightly regulated multi-step process. The key steps involved are: condensation of DNA, the cellular uptake of nanoparticle–DNA complexes, escape from degradation vesicles, intracellular movement or "trafficking," nuclear translocation and finally unpacking followed by translation. During this process of gene transfer the nanoparticle–DNA complexes have to bypass several barriers imposed by the physiological or metabolic processes of the cell. These barriers can be categorized as, *in vitro* and *in vivo* barriers. These barriers should be kept in mind before designing vectors for *in vitro* gene transfer, which can potentially be employed for *in vivo* applications. The present chapter cites the barriers encountered during process of gene delivery and suggested remedies.

Key words: enzymatic degradation, complex stability, internalization, endosomal escape, trafficking, nuclear localization, nucleases, endothelial barrier.

Published by Woodhead Publishing Limited, 2013

5.1 Introduction

The promise of gene therapy relies on the development of safe and efficient vectors for *in vivo* gene delivery. Although viral vectors are quite efficient for *in vivo* gene transfer, safety issues associated with them hamper their future use, rendering synthetic carriers as potential alternatives. Synthetic vectors such as cationic liposomes, polycations and nanoparticles are safer alternatives; however, their therapeutic application is hindered by low gene transfer efficiency. There is a need to perform mechanistic studies to identify the key rate-limiting steps in the non-viral gene delivery process. Further systematic and controlled studies are needed in order to elucidate the structure–function relationship. In this chapter, we explore the underlying mechanism of gene delivery and identify the key *in vitro* and *in vivo* barriers towards successful gene transfer.

5.2 Mechanism of gene delivery

Gene or DNA delivery to mammalian cells is a well-orchestrated multi-step process. The key steps involved are: condensation of DNA, the cellular uptake of nanoparticle–DNA complexes, escape from degradation vesicles, intracellular movement or "trafficking," nuclear translocation, and finally unpacking, followed by translation (Figure 5.1). The major rate-limiting step would depend upon the nature of the polymer or nanoparticles employed.

Cationic polymers electrostatically interact with the anionic DNA and condense it into compact particles in the range of 20–200 nm in diameter.[1–3] The condensation of DNA results from the association of the polycation around the DNA phosphate groups. At a specific N/P ratio, the DNA

Published by Woodhead Publishing Limited, 2013

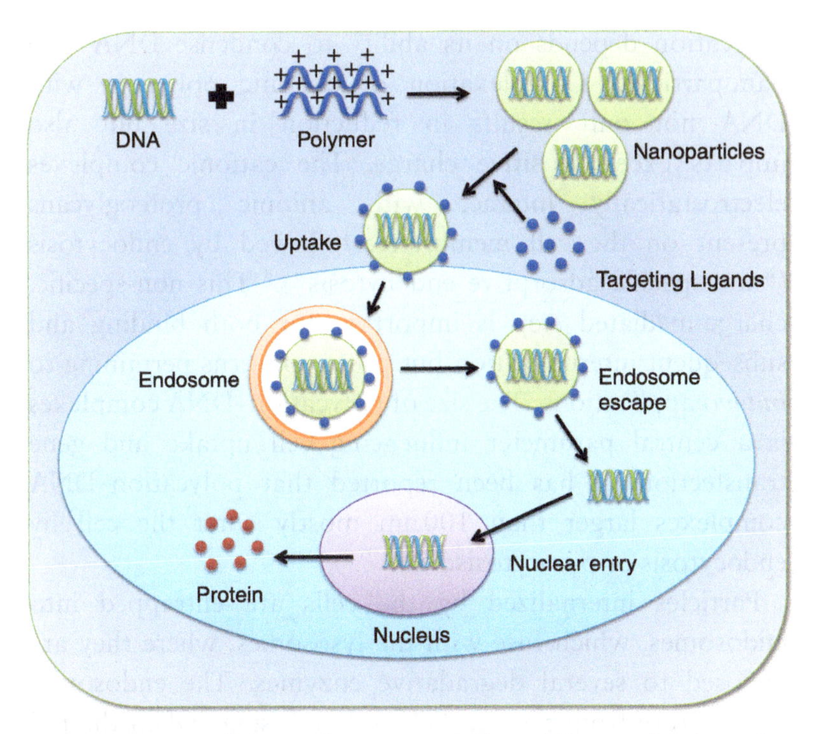

Figure 5.1 Schematic representation of the mechanism of gene delivery. It is a multi-step process that involves condensation of DNA, the cellular uptake of nanoparticle–DNA complexes, escape from degradation vesicles, intracellular movement or "trafficking," nuclear translocation, and finally unpacking, followed by translation

molecule undergoes localized bending or distortion, which facilitates the formation of rods, toroids, and spheroids.[1–3] Condensation of DNA by chitosan at N/P ratio between 3 and 8 leads to the formation of particles in the size range of 100–250 nm with a low polydispersity index (PDI).[4] We prepared chitosan/DNA nanoparticles by mixing a fixed amount of chitosan with DNA to obtain an N/P ratio of 5, which was shown to be most effective for chitosan 92-10 (MW-DDA).[5–7] The transfection efficiency of the

Published by Woodhead Publishing Limited, 2013

polycation depends on its ability to condense DNA into nanoparticles. Complexation of cationic polymers with DNA not only results in reduction in size but also imparts excess positive charge. The cationic complexes electrostatically interact with anionic proteoglycans present on the cell membrane, followed by endocytosis ("non-specific adsorptive endocytosis").[8] This non-specific, charge-mediated step is important for both binding and subsequent internalization but poses concerns pertaining to *in vivo* applications. The size of polycation–DNA complexes is a central parameter influencing cell uptake and gene transfection. It has been reported that polycation–DNA complexes larger than 100 nm mostly enter the cell by endocytosis or pinocytosis.[9]

Particles internalized by the cells are entrapped into endosomes, which fuse with the lysosomes, where they are exposed to several degradative enzymes. The endosomes first mature from the "early" to "late" stage when the pH drops from ~6 to ~5, and the late endosomes fuse with lysosomes.[10] The hydrolytic enzymes present inside the lysosomes would eventually lead to the degradation of complexes, resulting in inhibition of gene expression. However, cationic vectors that possess endosomolytic activity allow early escape of the polyplex from lysosomes, which accounts for enhanced transfection efficiency. The complexes must enter the nucleus to undergo transcription after their release from the endosomes into the cytosol. The transport of the complexes through the cytoplasm to the nucleus is poorly understood, but there is evidence that polycations protect DNA from cytosol nucleases, resulting in a higher probability of nuclear entry.[11] The tracking studies revealed that intact PEI/DNA polyplexes were found in the nucleus, which indicates that it may not be necessary for the polycation to separate from DNA prior to nuclear entry.[12–14]

Published by Woodhead Publishing Limited, 2013

Although the amount of complexes entering the nucleus is small, large amounts of complexes were observed in the perinuclear granular region in the case of *in vitro* transfection.[14] We observed a distinct relationship between transfection efficiency and polyplex dissociation rate.[15] The most efficient chitosans showed an intermediate stability and kinetics of dissociation, which occurred in synchrony with lysosomal escape. In contrast, a rapid dissociation before lysosomal escape was found for the inefficient low-DDA chitosan, whereas the highly stable and inefficient complex formed by a high-MW and high-DDA chitosan did not dissociate even after 24 h.[15] Release of DNA from the complexes to allow the transcription apparatus of the cell to access the DNA efficiently is the final stage in gene expression.

5.3 Barriers to gene delivery

Gene therapy has attracted much attention in industry and academia as potential therapeutics. However, the hidden potential of gene therapy cannot be tapped without evolution of strategies for controlled and targeted delivery of DNA. For an efficient transfection process, a cell usually internalizes 10^6 plasmid copies, of which only 10^2–10^4 can reach the nucleus for transgene expression.[16–18] This low efficiency of transfection is due to the inability of the delivery vector to cross the various barriers encountered right from the site of administration to localization in the cell nucleus. While undergoing the transfection process, nanoparticle–DNA complexes are exposed to several barriers that reduce their efficacy (Figure 5.2). These barriers can be categorized as *in vitro* and *in vivo* barriers.

Published by Woodhead Publishing Limited, 2013

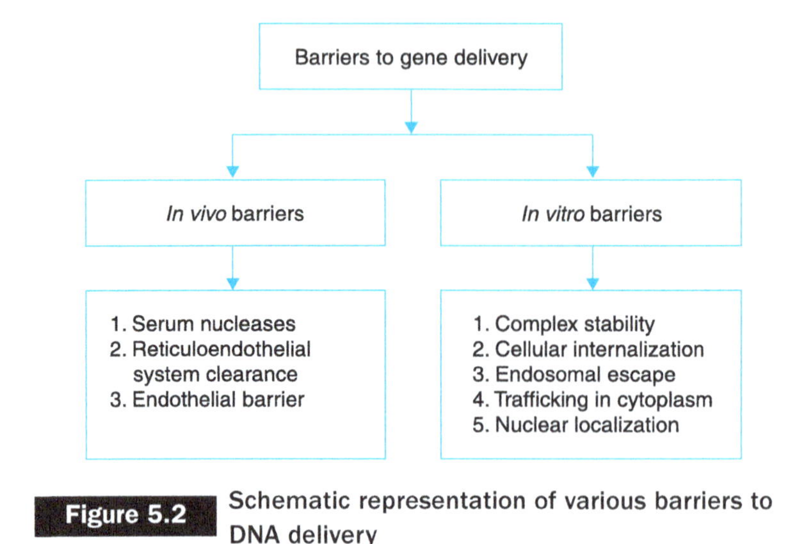

Figure 5.2 Schematic representation of various barriers to DNA delivery

5.3.1 In vitro *barriers*

The nanoparticle–DNA complexes have to bypass several challenges for efficient *in vitro* gene delivery efficacy. These barriers should be kept in mind before designing vectors for *in vitro* gene transfer, which can potentially be employed for *in vivo* applications as well.

5.3.1.1 Complex stability

Internalization of DNA by cells is inhibited by the large size and presence of anionic charge. Though several studies have proposed introduction of free DNA into cells by electroporation, or direct injection into target tissue, the therapeutic relevance of these techniques is limited.[19,20] Systemically administered DNA is rapidly degraded by nucleases; to address this issue DNA can be complexed with polycationic polymers. The size of the complexes thus formed depends on various factors such as concentration of DNA, type of polycationic polymer, pH, type of buffer, and N/P

Published by Woodhead Publishing Limited, 2013

ratio. Also, the size of polycationic polymer/DNA complexes is well correlated with MW of the polymer.[21] However, the colloidal stability of the vectors in extracellular compartments also turns out to be a major issue. The stability of complexes in the extracellular milieu depends on the chemical stability of DNA as well as the physical stability of the vector. Low colloidal stability of the polycationic polymers can be due to the presence of excess positive charge. However, to address this issue, polycationic polymers have been covalently linked with amphiphilic polymers such as PEG.[22,23] To characterize and quantify the physical stability, spherical nanoparticles of PLA and PMMA were prepared, in the size range 100–200 nm.[24] Stability analysis in salt solutions, biological fluids, serum and tissue homogenates by DLS revealed that PMMA nanoparticles were stable in all fluids, while PLA nanoparticles aggregated in gastric juice and spleen homogenate. The stability of chitosan nanoparticles cross-linked with tripolyphosphate (TPP) was investigated over a period of 1 month in terms of particle size and compactness.[25] Chitosan–TPP nanoparticles were prepared at different ionic strengths, chitosan chloride concentrations, and TPP-to-chitosan ratios. The particles were observed to be more stable when prepared and stored under saline conditions compared with water. With increase in TPP-to-chitosan ratios, size of nanoparticles increased due to higher concentration of chitosan, increased aggregation, and sedimentation of the particles with reduced colloidal stability of the nanoparticles.

5.3.1.2 Complex internalization

The mechanism of polyplex and nanoparticle internalization is a complex process and depends on several factors such as size, shape, type, composition, and surface charge of the nanoparticles; and also on the cell type and membrane composition involved.[26] Uptake of polymer/DNA complexes usually takes place by phagocytosis or by clathrin-mediated endocytosis. The polycationic complexes interact with cell surface heparin sulfate proteoglycans (HSPGs) present abundantly on all cell surfaces. Uptake of PEI/DNA complexes has been proposed to proceed by complex binding to transmembrane HSPGs, known as syndecans, which cluster into cholesterol-rich rafts on the cell surface.[27] This clustering triggers protein kinase C phosphorylation followed by binding of the syndecan to the actin skeleton through linker proteins. This binding leads to internalization of complex into the cell by phagocytosis. Several studies have reported the involvement of clathrin-coated pits in the internalization of polycationic vector/DNA complexes.[28,29]

A comparative study of endocytic pathways in various cell lines revealed that inhibition of caveolae-mediated endocytosis was more efficient in reducing the transfection of HeLa cells by PEI polyplexes than inhibition of clathrin-mediated endocytosis,[30] while in COS-7 and HUH-7 cells clathrin-mediated endocytosis was found to be most efficient. In HepG2 cells it was shown that the uptake of PLL/DNA complexes occurs either by clathrin-dependent endocytosis or by macropinocytosis.[31] Attachment of targeting moieties, such as folate receptor ligands, to complexes leads to internalization mediated by caveolae.[32] The effects of inhibitors of clathrin-mediated endocytosis (chlorpromazine and K+ depletion) and of caveolae-mediated uptake (filipin and genistein) on internalization of fluorescein isothiocyanate (FITC)-PLL-labeled 1,2-dioleoyl-3-trimethylammonium propane (DOTAP)/DNA lipoplexes and PEI/DNA polyplexes was investigated in A549 and HeLa cells. It was observed that lipoplex uptake proceeds only by clathrin-mediated endocytosis, while polyplexes are taken up by caveolae and clathrin-mediated endocytosis. Moreover, transfection

Published by Woodhead Publishing Limited, 2013

mediated by polyplexes was completely blocked by genistein and filipin but was unaffected by inhibitors of clathrin-mediated endocytosis.[33] In another study, the uptake of trisaccharide-substituted chitosan oligomers was compared with linear chitosan.[34] Linear chitosan polyplexes were internalized via clathrin-dependent as well as clathrin-independent pathways, whereas trisaccharide-substituted chitosan oligomer polyplexes were only taken up by clathrin-independent endocytosis.

5.3.1.3 Endosomal escape

The intracellular route of polyplexes is determined by the endocytic pathway. Polyplexes internalized by the clathrin-dependent pathway follow the classical endocytic pathway to endosomes and lysosomes, which could lead to degradation. Hence, lysosomal degradation of DNA poses one of the major obstacles for efficient expression of a therapeutic gene. Several mechanisms have been proposed for endosomal escape of vectors. One such hypothesis suggests physical disruption of the negatively charged endosomal membrane by direct interaction with the cationic polymer. Polyamidoamine (PAMAM) dendrimers and PLL have been shown to undergo endosomal escape by this mechanism.[35] Polycationic polymers such as PEI with ionizable amine groups use the "proton sponge" mechanism to promote the release of endocytosed polyplexes from the endosomes (Figure 5.3).[36–39] During the process of endosome maturation, the membrane-bound ATPase proton actively translocates protons from the cytosol into the endosomes, which results in the acidification of endosomal compartments and activation of hydrolytic enzymes. However, at this point the polycationic polymer will become protonated and resist the acidification of endosomes (Figure 5.3). To mask this

Published by Woodhead Publishing Limited, 2013

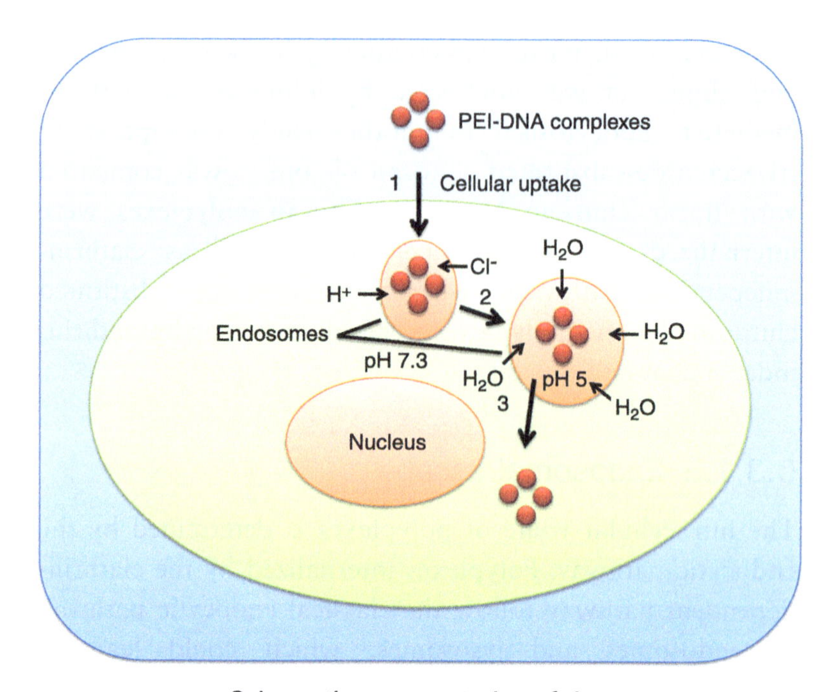

Figure 5.3 Schematic representation of the proton sponge mechanism, beginning with endocytosis of the cationic complexes (1), followed by acidification of endosomal compartments (2), which leads to increased osmotic pressure and finally rupture of endosomes (3)

action of polymers, more protons will continuously be pumped into the endosomes to achieve lower pH. The proton pumping action is followed by passive entry of chloride ions, increasing ionic concentration, followed by water influx. Ultimately, the increase in osmotic pressure causes swelling and rupture of endosomes, releasing their contents to the cytosol. Later, this hypothesis was validated by investigating the concentration of chloride ions, pH and the volume of endosomes after internalization of polyplexes composed of DNA and non-buffering PLL along with PEI and PAMAM.[40] In the case of PEI and PAMAM, high concentration of

Published by Woodhead Publishing Limited, 2013

chloride ion, volume expansion and membrane lysis were observed, which were absent with PLL. This buffering may protect DNA from degradation in the endosomal compartment during the maturation of the early endosomes to late endosomes and their subsequent fusion with the lysosomes.

However, polycationic polymers such as chitosan and PLL that lack pH buffering properties have been observed to possess poor transfection efficiency. To improve upon the gene delivery efficacy of such polymers, different functional moieties have been attached. Histidine, a commonly employed molecule, possesses buffering capacity due to the presence of an imidazole ring that has a pKa around 6 and can be protonated at acidic pH.[41,42] Cell penetrating peptides (CPPs) such as Glutamine-Alanine-Leucine-Alanine peptide (GALA) or fusogenic peptide (KALA) have also been incorporated in the polymer backbone to enhance buffering capacity. Exogenous additives, such as chloroquine and inactivated adenovirus, have also been exploited to promote endosomal escape and enhance gene delivery efficiency. Chloroquine neutralizes the acidic compartment and induces rupture of endocytic vesicles, thereby promoting escape of polymer/DNA complexes.[43] Once the polymer/DNA complexes are released into the cytoplasm, they must overcome additional barriers in the cytosol that hamper delivery of the complex into the nucleus of the host cell.

5.3.1.4 Trafficking in cytoplasm

Vectors that successfully escape endosomes in the peri-nuclear region of cells may still have to traverse distances of up to several micrometers through the molecularly crowded cytoplasm prior to reaching nuclear pores, the pathway into the nucleus. Hence, poor movement of complexes inside the

Published by Woodhead Publishing Limited, 2013

cytoplasm may critically hamper gene delivery efficacy. PEI/DNA nanoparticles have been observed to be efficiently transported to the peri-nuclear region of the cell by microtubules, leading to their rapid accumulation in that region within 30 minutes post-transfection.[44] Also, it was observed that microtubule-based motor proteins were responsible for the transport of the nanoparticles. PEGylation of 100-nm polystyrene nanoparticles was observed to increase the average nanoparticle diffusivities by 100% compared with non-PEGylated particles (timescale of 10 s) in live cells. Moreover, faster particle transport correlated with a marked decrease in the number of particles that underwent hindered transport, from 79.2% (unmodified) to 48.8% (PEGylated).[45]

5.3.1.5 Nuclear localization

For gene expression, it is a prerequisite for DNA to enter the nucleus and gain access to the cellular transcription machinery. In normal cells, DNA can enter the nucleus through nuclear pores, while in dividing cells DNA enters the nucleus during mitosis when nuclear envelope disassembly occurs, or it could transverse the nuclear envelope.[46] The movement of biomolecules across the nucleus takes place via the nuclear envelope, embedded in nuclear pore complexes (NPC).[47,48] The NPC consists of three distinct domains, a central domain, a nuclear and a cytoplasmic ring, made from ~ 50 different nucleoporin proteins.[49] Studies investigating the kinetics of nuclear transport have demonstrated that each NPC is capable of handling ~1000 translocation events per second, including both active and passive transport.[50] The nuclear pores allow passive diffusion of molecules less than 9 nm in diameter; however, the movement of molecules larger than 26 nm employs an ATP-dependent process triggered by re-organization of short peptide sequences.[51]

Published by Woodhead Publishing Limited, 2013

Significantly higher levels of gene expression have been observed in transfection employing DNA complexed with a polycationic vector than that of free plasmid DNA (pDNA), indicating that a positively charged vector may have a nuclear-localizing effect.[52] Several nuclear localization signals (NLS) have been incorporated into the polymers to promote the nuclear uptake of polymer/DNA complexes. Polycationic polymers have been observed to enhance DNA entry into the nucleus by association with nuclear material during mitosis when disassembly of the nuclear envelope occurs. Improved transfection efficiencies have been reported with polycationic polymers as compared with free DNA with transfection carried out in different stages of the cell cycle.[46,53] Transfection experiments with DNA complexed to polycationic polymers appear to be advantageous as compared with free DNA, independent of the mechanism of nuclear uptake.

5.3.2 In vivo *barriers*

Extracellular barriers to DNA delivery depend on the route of administration (e.g. intravenous, intranasal, intratracheal, subcutaneous, intratumor, intramuscular, or oral), which, in turn, depends upon the targeted disease. It is worth mentioning that the barriers encountered by native oligonucleotides, i.e. siRNA, antisense molecules, or aptamers, will be quite different from those encountered by oligonucleotides associated with various polymeric nanoparticles. Here we mention some of the common barriers encountered during *in vivo* DNA administration.

5.3.2.1 Degradation by serum nuclease

Systemically delivered free DNA or oligonucleotides (ODNs) are rapidly degraded by nucleases in serum, with a half-life

Published by Woodhead Publishing Limited, 2013

varying from a few minutes to several hours. The major enzymatic activity that occurs in the plasma is the 3′ exonuclease, although cleavage of inter-nucleotide bonds can also occur. Chemical modifications involving alteration in the ODN backbone have been observed to significantly improve the stability of ODN in the biological milieu. Replacement of the non-bridging oxygen of the phosphodiester backbone by sulfur resulted in the synthesis of phosphorothioate ODN with enhanced stability towards enzymatic degradation.[54] However, phosphorothioate ODN tends to bind non-specifically to proteins, thus causing toxicity. Other chemical modification includes replacement of the non-bridging oxygen by a methyl group which results in a methylphosphonate DNA, 2′-OH modifications, locked nucleic acids (LNA), peptide nucleic acids (PNA), morpholino compounds, and hexitol nucleic acids (HNA).[55,56] However, ODN associated with polymeric nanoparticles easily bypasses this barrier, as the condensed ODN is no longer available for 3′ nuclease binding followed by cleavage.

5.3.2.2 Clearance via the reticuloendothelial system

Another major obstacle is the renal clearance by the RES. Phagocytic cells of the RES, more specifically the Kupffer cells in the liver and splenic macrophages, can endocytose DNA as well as carriers used to deliver it.[57] Foreign particles such as DNA molecules in circulation are bound by opsonins, which consist of immunoglobulins, complement system proteins, and other serum proteins. These opsonized particles are recognized by a variety of receptors present on the cell surface of macrophages. Immunoglobulin G-opsonized particles are recognized by Fc receptors and complement-opsonized particles are internalized through complement

Published by Woodhead Publishing Limited, 2013

receptors, thereby leading to their degradation. However, surfaces of nanoparticles have been modified with hydrophilic polymers such as PEG, which reduces the adsorption of opsonins, followed by reduced clearance by phagocytosis.[58] The covalent coupling of polymers with PEG results in shielding the surface charge of polycations. PEG is a preferred candidate for such steric stabilization due to its charge neutrality and water solubility.

5.3.2.3 Endothelial barrier

The endothelial cells present on the surface of the vascular lumen present both a barrier and an advantage for oligonucleotide-based therapeutics. On one hand, they comprise a major cell type that is easily accessed by systemically administered drugs. It has been well established that endothelial cells have a profound role in several pathologies, such as tumors and blood pressure, and hence they appear to be a potential therapeutic target. On the other hand, when ODN needs to transit across the endothelium to be delivered to tissue parenchymal cells, it becomes a major obstacle. In most tissues the structure of the endothelia is tight, restricting egress of materials larger than 4 or 5 nm. Only organs and tissues with an irregular fenestration, such as the liver, spleen, bone marrow, and certain tumors, have endothelia with large meshes. Hence, nanoparticles may be useful for delivery to certain types of tumors and to normal tissues having fenestrated endothelia. In this regard, complex size plays an important role, since only small particles can pass through the fenestrated barriers known as enhanced permeability and retention (EPR). However, before cellular entry, the delivery system should be able to cross the plasma membrane barrier, composed of a variety of polysaccharides and proteins over the surface of cells that produce them.[59]

Published by Woodhead Publishing Limited, 2013

5.4 Conclusions

Although the issues pertaining to stability and efficacy have made reasonable progress, efficient intracellular delivery still remains a key issue. The problems are further aggravated when transition is made from *in vitro* to *in vivo* gene delivery. Ideally, the nanoparticle-based delivery system should avoid both interactions with plasma proteins and uptake by the macrophages. It is envisaged that use of hydrophilic polymers may be fruitful, but a considerable amount of attention needs to be paid to biologically based strategies. Hence, several factors should be considered at both cellular and organ levels in the quest towards design of efficient *in vivo* gene delivery strategies.

5.5 References

1. Dunlap, D.D., Maggi, A., Soria, M.R. and Monaco, L. (1997) Nanoscopic structure of DNA condensed for gene delivery. *Nucleic Acids Res*, **25**, 3095–101.
2. Golan, R., Pietrasanta, L.I., Hsieh, W. and Hansma, H.G. (1999) DNA toroids: stages in condensation. *Biochemistry*, **38**, 14069–76.
3. Liu, G., Molas, M., Grossmann, G.A., Pasumarthy, M., Perales, J.C., et al. (2001) Biological properties of poly-L-lysine-DNA complexes generated by cooperative binding of the polycation. *J Biol Chem*, **276**, 34379–87.
4. Mao, H.Q., Roy, K., Troung-Le, V.L., Janes, K.A., Lin, K.Y., et al. (2001) Chitosan-DNA nanoparticles as gene carriers: synthesis, characterization and transfection efficiency. *J Control Release*, **70**, 399–421.
5. Lavertu, M., Methot, S., Tran-Khanh, N. and Buschmann, M.D. (2006) High efficiency gene transfer

Published by Woodhead Publishing Limited, 2013

using chitosan/DNA nanoparticles with specific combinations of molecular weight and degree of deacetylation. *Biomaterials*, **27**, 4815–24.

6. Jean, M., Smaoui, F., Lavertu, M., Methot, S., Bouhdoud, L., et al. (2009) Chitosan-plasmid nanoparticle formulations for IM and SC delivery of recombinant FGF-2 and PDGF-BB or generation of antibodies. *Gene Ther*, **16**, 1097–110.
7. Nimesh, S., Thibault, M.M., Lavertu, M. and Buschmann, M.D. (2010) Enhanced gene delivery mediated by low molecular weight chitosan/DNA complexes: effect of pH and serum. *Mol Biotechnol*, **46**, 182–96.
8. Erbacher, P., Bettinger, T., Belguise-Valladier, P., Zou, S., Coll, J.L., et al. (1999) Transfection and physical properties of various saccharide, poly(ethylene glycol), and antibody-derivatized polyethylenimines (PEI). *J Gene Med*, **1**, 210–22.
9. Wolfert, M.A. and Seymour, L.W. (1996) Atomic force microscopic analysis of the influence of the molecular weight of poly(L)lysine on the size of polyelectrolyte complexes formed with DNA. *Gene Ther*, **3**, 269–73.
10. Luzio, J.P., Mullock, B.M., Pryor, P.R., Lindsay, M.R., James, D.E., et al. (2001) Relationship between endosomes and lysosomes. *Biochem Soc Trans*, **29**, 476–80.
11. Moret, I., Esteban Peris, J., Guillem, V.M., Benet, M., Revert, F., et al. (2001) Stability of PEI-DNA and DOTAP-DNA complexes: effect of alkaline pH, heparin and serum. *J Control Release*, **76**, 169–81.
12. Godbey, W.T., Barry, M.A., Saggau, P., Wu, K.K. and Mikos, A.G. (2000) Poly(ethylenimine)-mediated transfection: a new paradigm for gene delivery. *J Biomed Mater Res*, **51**, 321–8.

Published by Woodhead Publishing Limited, 2013

13. Godbey, W.T., Wu, K.K. and Mikos, A.G. (1999) Size matters: molecular weight affects the efficiency of poly(ethylenimine) as a gene delivery vehicle. *J Biomed Mater Res*, **45**, 268–75.
14. Bieber, T., Meissner, W., Kostin, S., Niemann, A. and Elsasser, H.P. (2002) Intracellular route and transcriptional competence of polyethylenimine-DNA complexes. *J Control Release*, **82**, 441–54.
15. Thibault, M., Nimesh, S., Lavertu, M. and Buschmann, M.D. (2010) Intracellular trafficking and decondensation kinetics of chitosan-pDNA polyplexes. *Mol Ther*, **18**, 1787–95.
16. Tachibana, R., Harashima, H., Ide, N., Ukitsu, S., Ohta, Y., et al. (2002) Quantitative analysis of correlation between number of nuclear plasmids and gene expression activity after transfection with cationic liposomes. *Pharm Res*, **19**, 377–81.
17. Felgner, P.L., Holm, M. and Chan, H. (1989) Cationic liposome mediated transfection. *Proc West Pharmacol Soc*, **32**, 115–21.
18. James, M.B. and Giorgio, T.D. (2000) Nuclear-associated plasmid, but not cell-associated plasmid, is correlated with transgene expression in cultured mammalian cells. *Mol Ther*, **1**, 339–46.
19. Potter, H. (1988) Electroporation in biology: methods, applications, and instrumentation. *Anal Biochem*, **174**, 361–73.
20. Mir, L.M., Bureau, M.F., Gehl, J., Rangara, R., Rouy, D., et al. (1999) High-efficiency gene transfer into skeletal muscle mediated by electric pulses. *Proc Natl Acad Sci USA*, **96**, 4262–7.
21. Tang, M.X. and Szoka, F.C. (1997) The influence of polymer structure on the interactions of cationic

Published by Woodhead Publishing Limited, 2013

polymers with DNA and morphology of the resulting complexes. *Gene Ther*, **4**, 823–32.

22. Choi, Y.H., Liu, F., Kim, J.S., Choi, Y.K., Park, J.S., et al. (1998) Polyethylene glycol-grafted poly-L-lysine as polymeric gene carrier. *J Control Release*, **54**, 39–48.
23. Oupicky, D., Konak, C., Ulbrich, K., Wolfert, M.A. and Seymour, L.W. (2000) DNA delivery systems based on complexes of DNA with synthetic polycations and their copolymers. *J Control Release*, **65**, 149–71.
24. Lazzari, S., Moscatelli, D., Codari, F., Salmona, M., Morbidelli, M., et al. (2012) Colloidal stability of polymeric nanoparticles in biological fluids. *J Nanopart Res*, **14**, 1–10.
25. Jonassen, H., Kjoniksen, A.L. and Hiorth, M. (2012) Stability of chitosan nanoparticles cross-linked with tripolyphosphate. *Biomacromolecules*, **13**, 3747–56.
26. Ruponen, M., Honkakoski, P., Tammi, M. and Urtti, A. (2004) Cell-surface glycosaminoglycans inhibit cation-mediated gene transfer. *J Gene Med*, **6**, 405–14.
27. Kopatz, I., Remy, J.S. and Behr, J.P. (2004) A model for non-viral gene delivery: through syndecan adhesion molecules and powered by actin. *J Gene Med*, **6**, 769–76.
28. Zuhorn, I.S., Visser, W.H., Bakowsky, U., Engberts, J.B. and Hoekstra, D. (2002) Interference of serum with lipoplex-cell interaction: modulation of intracellular processing. *Biochim Biophys Acta*, **1560**, 25–36.
29. Zuhorn, I.S., Kalicharan, R. and Hoekstra, D. (2002) Lipoplex-mediated transfection of mammalian cells occurs through the cholesterol-dependent clathrin-mediated pathway of endocytosis. *J Biol Chem*, **277**, 18021–8.
30. von Gersdorff, K., Sanders, N.N., Vandenbroucke, R., De Smedt, S.C., Wagner, E., et al. (2006) The

Published by Woodhead Publishing Limited, 2013

internalization route resulting in successful gene expression depends on both cell line and polyethylenimine polyplex type. *Mol Ther*, **14**, 745–53.

31. Goncalves, C., Mennesson, E., Fuchs, R., Gorvel, J.P., Midoux, P., et al. (2004) Macropinocytosis of polyplexes and recycling of plasmid via the clathrin-dependent pathway impair the transfection efficiency of human hepatocarcinoma cells. *Mol Ther*, **10**, 373–85.
32. Hofland, H.E., Masson, C., Iginla, S., Osetinsky, I., Reddy, J.A., et al. (2002) Folate-targeted gene transfer in vivo. *Mol Ther*, **5**, 739–44.
33. Rejman, J., Bragonzi, A. and Conese, M. (2005) Role of clathrin- and caveolae-mediated endocytosis in gene transfer mediated by lipo- and polyplexes. *Mol Ther*, **12**, 468–74.
34. Garaiova, Z., Strand, S.P., Reitan, N.K., Lelu, S., Storset, S.O., et al. (2012) Cellular uptake of DNA-chitosan nanoparticles: The role of clathrin- and caveolae-mediated pathways. *Int J Biol Macromol*, **51**, 1043–51.
35. Zhang, Z.Y. and Smith, B.D. (2000) High-generation polycationic dendrimers are unusually effective at disrupting anionic vesicles: membrane bending model. *Bioconjugate Chem*, **11**, 805–14.
36. Boussif, O., Lezoualc'h, F., Zanta, M.A., Mergny, M.D., Scherman, D., et al. (1995) A versatile vector for gene and oligonucleotide transfer into cells in culture and in vivo: polyethylenimine. *Proc Natl Acad Sci USA*, **92**, 7297–301.
37. Thomas, M., Lu, J.J., Ge, Q., Zhang, C., Chen, J., et al. (2005) Full deacylation of polyethylenimine dramatically boosts its gene delivery efficiency and specificity to mouse lung. *Proc Natl Acad Sci USA*, **102**, 5679–84.

Published by Woodhead Publishing Limited, 2013

38. Thomas, M. and Klibanov, A.M. (2002) Enhancing polyethylenimine's delivery of plasmid DNA into mammalian cells. *Proc Natl Acad Sci USA*, **99**, 14640–5.
39. Akinc, A., Thomas, M., Klibanov, A.M. and Langer, R. (2005) Exploring polyethylenimine-mediated DNA transfection and the proton sponge hypothesis. *J Gene Med*, 7, 657–63.
40. Sonawane, N.D., Szoka, F.C., Jr. and Verkman, A.S. (2003) Chloride accumulation and swelling in endosomes enhances DNA transfer by polyamine-DNA polyplexes. *J Biol Chem*, **278**, 44826–31.
41. Fajac, I., Allo, J.C., Souil, E., Merten, M., Pichon, C., et al. (2000) Histidylated polylysine as a synthetic vector for gene transfer into immortalized cystic fibrosis airway surface and airway gland serous cells. *J Gene Med*, **2**, 368–78.
42. Pichon, C., Goncalves, C. and Midoux, P. (2001) Histidine-rich peptides and polymers for nucleic acids delivery. *Adv Drug Deliv Rev*, **53**, 75–94.
43. Erbacher, P., Roche, A.C., Monsigny, M. and Midoux, P. (1996) Putative role of chloroquine in gene transfer into a human hepatoma cell line by DNA/lactosylated polylysine complexes. *Exp Cell Res*, **225**, 186–94.
44. Suh, J., Wirtz, D. and Hanes, J. (2003) Efficient active transport of gene nanocarriers to the cell nucleus. *Proc Natl Acad Sci U S A*, **100**, 3878–82.
45. Suh, J., Choy, K.L., Lai, S.K., Suk, J.S., Tang, B.C., et al. (2007) PEGylation of nanoparticles improves their cytoplasmic transport. *Int J Nanomedicine*, **2**, 735–41.
46. Wilke, M., Fortunati, E., van den Broek, M., Hoogeveen, A.T. and Scholte, B.J. (1996) Efficacy of a peptide-based gene delivery system depends on mitotic activity. *Gene Ther*, 3, 1133–42.

47. Harel, A. and Forbes, D.J. (2004) Importin beta: conducting a much larger cellular symphony. *Mol Cell*, **16**, 319–30.
48. Wente, S.R. (2000) Gatekeepers of the nucleus. *Science*, **288**, 1374–7.
49. Ryan, K.J. and Wente, S.R. (2000) The nuclear pore complex: a protein machine bridging the nucleus and cytoplasm. *Curr Opin Cell Biol*, **12**, 361–71.
50. Ribbeck, K. and Gorlich, D. (2001) Kinetic analysis of translocation through nuclear pore complexes. *EMBO J*, **20**, 1320–30.
51. Ryan, K.J. and Wente, S.R. (2000) The nuclear pore complex: a protein machine bridging the nucleus and cytoplasm. *Curr Opin Cell Biol*, **12**, 361–71.
52. Pouton, C.W. (2001) Polymeric materials for advanced drug delivery. *Adv Drug Deliv Rev*, **53**, 1–3.
53. Vacik, J., Dean, B.S., Zimmer, W.E. and Dean, D.A. (1999) Cell-specific nuclear import of plasmid DNA. *Gene Ther*, **6**, 1006–14.
54. Krieg, A.M. and Stein, C.A. (1995) Phosphorothioate oligodeoxynucleotides: antisense or anti-protein? *Antisense Res Dev*, **5**, 241.
55. Kurreck, J. (2003) Antisense technologies. Improvement through novel chemical modifications. *Eur J Biochem*, **270**, 1628–44.
56. Grunweller, A., Wyszko, E., Bieber, B., Jahnel, R., Erdmann, V.A., et al. (2003) Comparison of different antisense strategies in mammalian cells using locked nucleic acids, 2′-O-methyl RNA, phosphorothioates and small interfering RNA. *Nucleic Acids Res*, **31**, 3185–93.
57. Alexis, F., Pridgen, E., Molnar, L.K. and Farokhzad, O.C. (2008) Factors affecting the clearance and biodistribution of polymeric nanoparticles. *Mol Pharm*, **5**, 505–15.

Published by Woodhead Publishing Limited, 2013

58. van Vlerken, L.E., Vyas, T.K. and Amiji, M.M. (2007) Poly(ethylene glycol)-modified nanocarriers for tumor-targeted and intracellular delivery. *Pharm Res*, **24**, 1405–14.
59. Li, W.J. and Szoka, F.C. (2007) Lipid-based nanoparticles for nucleic acid delivery. *Pharm Res*, **24**, 438–49.

Published by Woodhead Publishing Limited, 2013

6

Targeted gene delivery mediated by nanoparticles

DOI: 10.1533/9781908818645.113

Abstract: Success of gene delivery depends on transport of DNA to desired tissues or organs. Target specific gene delivery employing nanoparticles can be achieved by decorating with various targeting ligands such as peptides, antibodies, and sugar molecules. Targeting strategies can be broadly classified as passive and active targeting. For passive targeting naïve nanoparticles are administered leading to biodistribution within the body, followed by delivery of payload to various organs. However, active targeting employs use of targeting moieties specific for different tissues. The present chapter accounts for various strategies employed to achieve gene delivery to desired tissues/organs.

Key words: targeting, passive targeting, active targeting, ligands, antibodies, transferrin, peptides, cholesterol.

6.1 Introduction

One of the major hurdles towards successful clinical application of gene delivery is the lack of specificity for target

Published by Woodhead Publishing Limited, 2013

cells. Several attempts have been made to achieve target-specific gene delivery employing nanoparticles decorated with various targeting ligands such as peptides, antibodies, and sugar molecules. The uptake of ligand-modified nanoparticles occurs via interaction of ligands with the specific cell receptor, which further favors internalization. Targeted gene delivery not only allows the desired therapeutic effect to be achieved at low doses but also minimizes the possible adverse effects. Targeting approaches can be broadly classified into passive and active targeting.

6.1.1 Passive targeting

Passive targeting of nanoparticles can be achieved by biodistribution of naïve nanoparticles within the body. After intravenous administration, naïve nanoparticles are rapidly cleared from systemic circulation by opsonization and macrophage engulfment or accumulate in the liver and spleen.[1,2] Hence, this rapid clearance can be exploited for the treatment of hepatic disorders such as leishmaniasis, a parasitic disease, or for targeting of accumulated macrophages in atherosclerosis.[3,4] Studies have evidenced accumulation of particles up to 400 nm by passive targeting to tumors.[5] This could be achieved due to the ability of these particles to leach into the diseased tissue through the leaky vasculature network commonly observed in tumorigenesis, a phenomenon called the EPR effect (Figure 6.1).[5] When tumor cells undergo multiplication, cluster together and reach a size of 2–3 mm, they induce angiogenesis to meet the ever-increasing nutrition and oxygen demands of the growing tumor.[6] This neovasculature significantly differs from that of normal tissues in microscopic anatomical architecture.[7]

Published by Woodhead Publishing Limited, 2013

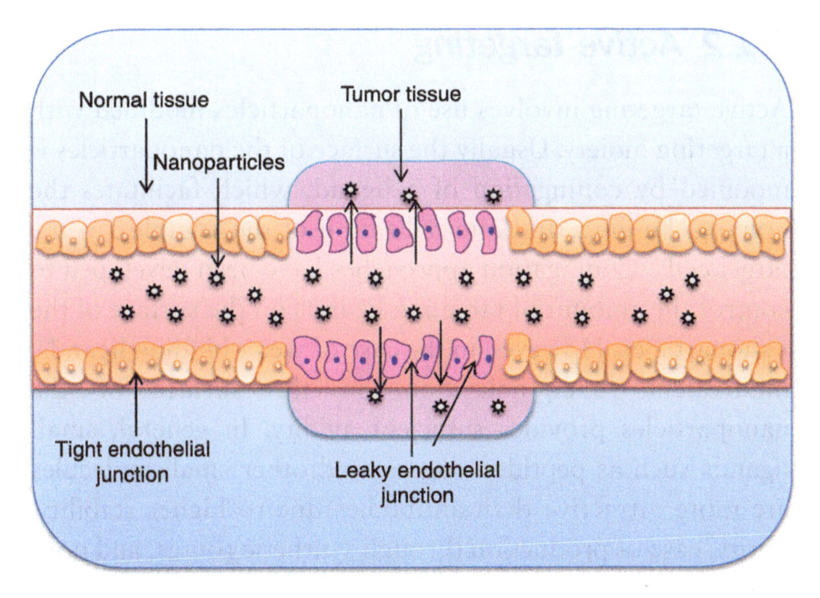

Figure 6.1 Schematic representation of tumor targeting by nanoparticles via EPR effect

The blood vessels in the tumor are irregular in shape, dilated, leaky, or defective, and the endothelial cells are poorly aligned or disorganized with large fenestrations (Figure 6.1). Moreover, the perivascular cells and the basement membrane, or the smooth-muscle layer, are frequently absent or abnormal in the vascular wall. Tumor vessels have a wide lumen, whereas tumor tissues have poor lymphatic drainage.[7–11] These anatomical defects, along with functional abnormalities, result in extensive leakage of blood plasma components, such as macromolecules, nanoparticles, and lipid particles, into the tumor tissue. Furthermore, the slow venous return in tumor tissue and the poor lymphatic clearance results in accumulation of macromolecules in the tumor, whereas extravasation into tumor interstitium continues.

Published by Woodhead Publishing Limited, 2013

6.1.2 Active targeting

Active targeting involves use of nanoparticles modified with a targeting moiety. Usually the surface of the nanoparticles is modified by conjugation of a ligand, which facilitates the homing, binding, and internalization of the complex to the target cells. Conjugation approaches have been developed to control the amount of targeting ligands on the surface of the nanoparticles. However, in the case of weak binding ligands, multivalent functionalization on the surface of the nanoparticles provides sufficient avidity. In general, small ligands such as peptides, sugars, and other small molecules are more attractive than antibodies due to higher stability, purity, ease of production through synthetic routes, and non-immunogenicity.

6.2 Approaches for targeted gene delivery

Cellular targeting involves efficient internalization of receptors upon binding to their ligands, followed by rapid recycling on the cell surface to allow repeated targeting. Two different strategies have been proposed for receptor-mediated targeting. The first approach consists of targeting the tumor micro-environment, including the extracellular matrix or surface receptors on endothelial cells of the tumor blood vessels, which is usually effective for the delivery of immune induction or anti-angiogenesis molecules. The second approach involves targeting receptors expressed on the surface of tumor cells for intracellular delivery of cytotoxic agents or signal-pathway inhibitors. Nanoparticles decorated with ligands towards the extracellular markers of transmembrane tumor antigens are generally taken up by

Published by Woodhead Publishing Limited, 2013

cancer cells via receptor-mediated endocytosis, leading to efficient intracellular delivery of therapeutic payloads. Although it is difficult to compare the efficacies of two approaches, recently receptor-targeted nanoparticles have shown exciting results. Hence, a key parameter is to design ligands for efficient gene delivery to the cell surface receptors. Cells undergoing tumorigenesis express a large number of proteins, which can be exploited to modulate delivery system for target specificity. Various strategies and advances for targeted gene delivery are discussed below in detail.

6.2.1 Antibodies

Ever since the discovery of the potential application of monoclonal antibodies (mAbs), they have been the preferred class of targeting molecules. Recent progress towards development of antibodies focuses on chimeric, humanized, and fully humanized derivatives to mask their immunogenicity. Antibodies have been conjugated with various polymeric nanoparticles for targeted delivery to tumors and other diseases. In a recent study, biodegradable cationic PHPA-PEI was complexed with pDNA to form nanoparticles, followed by conjugation of 9B9mAb, an anti-epidermal growth factor receptor (EGFR) mAb.[12] The size of the complex was observed to be around 300 nm at its optimal weight ratio, and it was efficiently internalized by SMMC–7721 cells. Further, the cytotoxicity of the complexes was found to be much lower than that of PEI 25 kD in SMMC–7721, HepG2, Bel–7404 and COS–7 cell lines. The complex efficiently delivered therapeutic AChE gene, which resulted in significantly improved anti-tumor effect on tumor-bearing nude mice. In another study, single domain antibody, also known as nanobody (anti-MUC1), was covalently conjugated to the distal end of PEG_{3500} in PEG_{3500}-PEI (25 kDa) conjugates

Published by Woodhead Publishing Limited, 2013

and used to deliver plasmids coding a transcriptionally targeted truncated-Bid (tBid) killer gene under the control of the cancer-specific MUC1 promoter.[13] The generated polyplexes were quite suitable for transfection and effectively increased the level of Bid/tBid expression, in both MUC1 over-expressing caspase 3-deficient (MCF7 cells) and caspase 3-positive (T47D and SKBR3) tumor cell lines and, concomitantly, induced considerable cell death. Neither transgene expression nor cell death occurred when the MUC1 promoter was replaced with the central nervous system (CNS)-specific synapsin I promoter.[13]

To improve biocompatibility of the gene vector and to avoid its being cleared by the host RES, PEI was modified with PEG followed by conjugation with G250 mAb.[14] G250 is one of the most extensively studied mAbs associated with the HeLa cell line.[15] The highest transfection efficiency was observed in HeLa cells as determined by flow cytometry after transfection with the gene encoding EGFP. Inhibition of surface antigen on the cell membrane of HeLa cells, by incubation with free G250 mAb, or by downregulation of G250 expression by siRNA transfection, resulted in a remarkable decrease in transfection efficiency. Further *in vivo* study with the complexes exhibited high transfection efficiency in tumors with no obvious toxicity. Another group developed a PEI (25 kDa)-based gene delivery system, with two model mAbs with different well-characterized antigen specificities: the mouse anti-human IgG1 mAb AS02 recognizing human CD90 (hThy–1), which is expressed on human fibroblasts, and the humanized anti-Her–2/neu mAb Trastuzumab, employed for the treatment of Her–2/neu-positive breast cancer. Efficacy and selectivity of gene delivery were evaluated for covalent mAb–PEI conjugates coupled with N-succinimidyl–3-(2-pyridyldithio)proprionate (SPDP) or N-succinimidyl–4-(maleimidomethyl)-cyclohexancarboxylate (SMCC) as a

Published by Woodhead Publishing Limited, 2013

newly introduced coupling reagent, and non-covalent complexes of mAb with IBFB 110001 coupled to PEI. While mAb-PEI conjugates coupled with SPDP resulted in antigen-non-specific EGFP expression, conjugates coupled with SMCC or IBFB 110001 enabled antigen -specific gene delivery. Han et al. also reported efficient and specific delivery of PEI conjugated with antibodies (PEI-anti-MMP–2) against MMP–2, a surface marker protein on cancer cells.[16]

To deliver genes to Her2-expressing cell lines, conjugates consisting of PEG (2000 Da)-graft-PEI (25 kDa) covalently coupled to Trastuzumab via SPDP were synthesized.[17] Conjugate complex sizes were observed to be in the range 130–180 nm with zeta potentials at different N/P ratios close to neutral. Complexes showed efficient binding and uptake by Her2-positive SK-BR–3 cells, but negligible binding and uptake in Her2-negative OVCAR–3 cells. Further, the reporter gene expression using targeted complexes in SK-BR–3 cells was up to seven-fold higher than that of unmodified PEG–PEI complexes. Moreover, inhibition experiments with free Trastuzumab showed a significant decrease in reporter gene expression using SK-BR–3 cells but no decrease using OVCAR–3 cells, strongly supporting a specific Her2-receptor-mediated uptake mechanism. In an earlier study, polyplexes of DNA–linear PEI conjugated to Trastuzumab were prepared to target breast cancer cell lines.[18] The polyplexes showed significantly greater (up to 20-fold) transfection activity than non-derivatized PEI-based polyplexes in the HER2 over-expressing SK-BR–3 cells, in contrast to low HER2-expressing MDA-MB–231 breast cancer cells. Further, the transfection efficiency of polyplexes was retained in serum-containing medium.

Although enormous progress has been made towards development of mAb-conjugated nanoparticles, it still encounters several barriers. Due to their large size and

Published by Woodhead Publishing Limited, 2013

complexity, there is need for significant manipulations for efficient gene delivery. Additionally, antibodies are far more expensive to generate than small-molecule drugs. Furthermore, the hydrodynamic size of antibodies is ~20 nm, which could contribute an increase in size of nanoparticles by up to 40 nm in proportion to the number of antibodies functionalized on them.[19,20]

6.2.2 Transferrin

Cellular uptake and delivery of iron takes place via interaction and internalization of iron-loaded transferrin (Tf) mediated by the transferrin receptors (TFR).[21] Transferrin is a monomeric glycoprotein that can transport one (mono-ferric) or two (di-ferric) iron atoms. It is conjugated to polyplexes by reductive amination of its carbohydrate residue, enabling its application as a targeting moiety to deliver genes, especially to tumor cells. Cell surface distribution of TFR is known to increase in several types of tumors, thereby providing a cell surface receptor which can be targeted. Internalization of TFR during endocytosis consists of recycling of TFR to the cell surface without its being degraded. In contrast to antibodies, transferrin is less expensive and commercially available in large quantities. Incorporation of transferrin into PEI/DNA polyplexes has been shown to enhance cellular internalization via receptor-mediated endocytosis and deliver genes specifically targeted to tumor tissue *in vivo*, even at low N/P ratios.[22–25] At higher grafting densities, transferrin exhibits effective charge shielding and therefore provides both polyplex protection and targeting. The most efficient gene delivery into tumors was achieved with vectors prepared by complexing pDNA with a blend of transferrin-tagged PEG–linear polyethylenimine (LPEI) copolymer, as compared with PEG–LPEI copolymer and LPEI itself. The transgene

Published by Woodhead Publishing Limited, 2013

expression was enhanced over equivalent systems derived from branched polyethylenimine (BPEI), transferrin-shielded particles, and transferrin-decorated PEG–BPEI–DNA complexes.[23,24]

The intratumoral injection of 188Re-labeled Tf-PEI (188Re-Tf-PEI) conjugate was studied for its efficacy in radionuclide therapy against tumor cells.[26] Human Burkitt's lymphoma xenografts were established in nude mice prior to intratumoral injection. Increased retention of 188Re was observed in mice treated with the 188Re-Tf-PEI. Tumors in these mice demonstrated extensive necrosis without widespread leakage of the radionuclide to neighboring tissues.[26] Further, to understand the interplay of factors which enhance the transcytosis of DNA nanoparticles across an *in vitro* model of M-cells, and to compare transport mediated by M-cells versus normal epithelial cells, chitosan/DNA nanoparticles were employed.[27] The nanoparticle transport through the M-cell co-culture model was observed to be five-fold that of the intestinal epithelial monolayer, with at least 80% of the chitosan–DNA nanoparticles internalized in the first 30 min. Moreover, incorporation of Tf into nanoparticles enhanced transport through both models by three to five-fold.

In a recent study, conjugation of Tf to PEG–PEI resulted in an efficient and safe vector for DNA delivery.[28] The nanocomplexes showed improved transfection efficiency in Jurkat cells, which are known to express elevated numbers of Tf receptors and reduced cytotoxicity as compared with PEI complex. In an earlier study, conjugates of the BPEI (25 kDa) with Tf were prepared to condense and deliver DNA.[29] The complexes yielded higher β-galactosidase gene expression than native PEI, in Jurkat cells. Transferrin-conjugated PEGylated polycyanoacrylate nanoparticles with a size <200 nm were prepared to encapsulate and deliver

Published by Woodhead Publishing Limited, 2013

pDNA.[30] *In vitro* studies showed a higher degree of cell attachment for these nanoparticles at 4°C compared to those without ligand in human K562 tumor cells overexpressing the TFR. However, the presence of free Tf significantly decreased the binding of nanoparticles to the cells, suggesting that the binding and uptake of the ligand was receptor-specific.[30]

6.2.3 Small peptides

Small peptide sequences are attractive targeting molecules due to their smaller size, lower immunogenicity, high stability, and ease of manipulation. The development of peptide phage libraries (~10^{11} different peptide sequences), bacterial peptide display libraries, plasmid peptide libraries, and new screening technologies have made their selection much easier, contributing to their popularity as targeting ligands. The expression of vitronectin receptor $\alpha v\beta 3$ is highly upregulated in tumor vascular cells, in contrast to minimal expression in resting or normal blood vessels.[31] The tripeptide sequence RGD, found at the active site of vitronectin, binds to $\alpha v\beta_3$ and almost half of the 22 known integrins.[32]

In one of the earlier studies, a variety of RGD–PEI conjugates with different degrees of substitution, with or without PEG spacer, were synthesized and used to deliver pDNA to Mewo human melanoma cells and A549 human lung carcinoma cells.[33] Coupling of RGD without a PEG spacer improved transfection efficiency of PEI in integrin-expressing Mewo cells by one to two orders of magnitude, especially at low N/P ratios. Disulfide-containing PEI (SS-PEI)/DNA complexes were ionically coupled to RGD and investigated for transfection.[34] *In vitro* transfection experiments revealed that SS-PEI exhibited comparable transfection efficiency, but reduced cytotoxicity, in

Published by Woodhead Publishing Limited, 2013

comparison with PEI (25 kDa). The transfection efficiency of complexes in HeLa cells was significantly reduced with the increase in content of RGD peptide, whereas it remained unaltered in 293T cells.[34] Hyaluronic acid, a natural anionic mucopolysaccharide, was used to coat PEI–Poly(γ-benzyl L-glutamate) (PBLG)/DNA complexes.[35] Coating the complexes with HA and HA–RGD resulted in lower surface charge and only slightly bigger size than the naked PEI–PBLG/DNA. Further, HA/PEI–PBLG/DNA complexes showed only slightly lower transfection efficiency compared with naked PEI–PBLG/DNA, while the transfection efficiency of HA–RGD/PEI–PBLG/DNA was 9.7 times that of HA/PEI–PBLG/DNA due to the RGD target binding affinity to the receptors on HeLa cells.[35]

TAT oligopeptide has been employed as a model CPP, by covalent coupling to the polymeric nanoparticles.[36] The basic domain of TAT peptide (RKKRRQRRR) is comprised of arginine and lysine amino acid residues that can play an important role in the translocation across biological membranes, due to strong cell adherence that is independent of receptors and temperature.[20] TAT has been shown to destabilize the lipid bilayer of the cell membrane through an energy-independent pathway carrying hydrophilic or macromolecules as large as several hundred nanometers in size across the plasma membrane.[21] RGD or HIV-1 TAT peptides were attached to PEI/DNA nanocomplexes via PEG spacer molecules and employed for gene delivery to undifferentiated and differentiated SH-SY5Y cells.[36] Both RGD and TAT improved the cellular uptake of gene vectors and enhanced the gene transfection efficiency of primary neurons up to 14-fold. RGD functionalization not only led to receptor-targeted delivery but also resulted in a significant increase in vector escape from endosomes.[36] TAT has been covalently coupled to PEI (25 kDa) through a

Published by Woodhead Publishing Limited, 2013

heterobifunctional PEG spacer, resulting in a TAT–PEG–PEI conjugate, and used for transfection in A549 cells and in mice after intratracheal instillation.[37] Although the luciferase expression in A549 cells was ten-fold less for TAT–PEG–PEI conjugate than for PEI, significantly higher transfection efficiencies were detected with TAT–PEG–PEI in mice.

Brain-derived neurotrophic factor (BDNF) is a protein of the neurotrophin class, a group of nerve growth factors that support the growth, differentiation, and survival of neuronal cells. Two neurotrophin receptors to which BDNF is capable of binding exist on the cell surface: the low-affinity neurotrophin receptor p75 and the high-affinity TrkB.[38,39] Therapeutic use of BDNF, e.g. in the treatment of neurodegenerative diseases, is, however, restricted because of its molecular size, short half-life, and side effects. Hence, to mimic the actions of BDNF, low-MW TrkB-binding peptides have been developed.[38] Chitosan/pDNA nanoparticles were functionalized by coupling fluorescent dye and TrkB-binding peptides on the particle surface and tested in TrkB-positive murine transformed monocyte/macrophage cells (RAW 264).[40] *In vitro*, binding and uptake studies revealed that TrkB-peptide-functionalized nanoparticles bound to cells more effectively than nanoparticles functionalized with a control peptide. Furthermore, the length of the PEG spacer arm of the amine-to-sulfhydryl cross-linker used in the functionalization was also found to positively correlate with the cellular attachment efficiency.

The nuclear membrane poses one of the major barriers to the delivery and expression of exogenous genes, and hence several attempts have been made to overcome this barrier. Among various strategies, the use of NLS peptides for non-viral gene transfer has been widely investigated.[41–43] NLS sequences are typically less than 12 residues in length and rich in basic amino acids, imparting a net positive charge and

Published by Woodhead Publishing Limited, 2013

attachment to larger molecules, which leads to recognition by cytoplasmic transport receptors and mediates nuclear uptake. The most studied NLS sequences are peptides derived from viruses like Tat (transactivating) protein or Antennapedia homeodomain protein-derived ones, but arginine/lysine-rich NLS, such as for the SV40 large T antigen (PKKKRKV), seem to be far more efficient than those peptides. The chitosan/DNA complexes containing NLS were observed to increase transfection efficiencies in a NLS-dose dependent manner in HeLa cells, compared with the control.[44] The highest transfection efficiency was observed in chitosan/DNA complex at the weight ratio of 8 with 120 μg NLS, and was 74-fold higher than that in the cells transfected with chitosan/DNA complex. However, the cytotoxicity of the NLS/chitosan/DNA complexes also increased as the amount of the peptide increased, with ~ 80% cell viability at the most effective peptide concentration.

6.2.4 Other targeting ligands

Several other targeting ligands, such as folate, carbohydrates and cholesterol, have been reported as targeting moieties. Folate receptors (FRs) are over-expressed on the cell surfaces in many different types of human cancers, including ovarian, lung, breast, brain, colon, and kidney.[45] Use of folic acid as a targeting ligand for FRs bears several advantages: (a) a small targeting ligand, which favors efficient pharmacokinetic properties of the folate conjugates and reduced immunogenicity; (b) easy availability and inexpensive; (c) ease of chemical manipulation; (d) possess high affinity for FRs; (e) the receptor–ligand complex can be induced to internalize via endocytosis; and (f) high occurrence of FR over-expression among human tumors. Therefore, folate presents an attractive target for tumor-selective delivery.

Published by Woodhead Publishing Limited, 2013

Folate-mediated targeting was induced by conjugating PEG–folate on histidine-modified chitosan (15 kDa) and high degree of quaternization.[46] The *in vitro* transfection efficiency in KB cell line, which over-expresses FRs, in the presence of 10% FBS was found to be comparable to the control. Further, the chitosan derivative promoted cell growth by up to 139% as compared with control under normal cell culture conditions. Chitosan (50 kDa) was conjugated with folate, followed by nanoparticle preparation using a coacervation process.[47] The intravenous administration of nanoparticles in the right posterior paw of normal and arthritic rats demonstrated 5 to 12 times more fluorescence intensity of red fluorescent protein (DsRed) in the right soleus muscle and in the right gastro muscle than other tissue sections. Furthermore, high expression of β-galactosidase gene was observed with folate/chitosan/pDNA nanoparticles in the soleus muscle.[47]

Asialoglycoprotein receptor (ASGPR) is exclusively found in hepatocytes, where it is located at the basolateral membrane and therefore faces the bloodstream. Human ASGPR is a hetero-oligomer which is composed of two homologous subunits (46 and 50 kDa) and is present at a high density of 500,000 receptors per cell, and retained on several human hepatoma cell lines. ASGPR possesses very high affinity (K_D in the nanomolar range) for tri- and tetra-antennary N-linked sugar side chains with terminal galactose residues. Several studies have reported use of a combination of polymeric gene carriers with ligands such as galactose, lactose, and apoprotein E.

Chitosan was modified with a lactobionic acid-bearing galactose group for hepatocyte targeting, and GC grafted with dextran (GCD) to improve stability in water.[48] Efficient transfection was observed with GCD/DNA complexes only into Chang liver cells and HepG2 having ASGPR, suggesting

Published by Woodhead Publishing Limited, 2013

a specific interaction of ASGPR on cells and galactose ligands on chitosan. Later, dextran was replaced with PEG, which resulted in high transfection efficiency in HepG2 cells.[49] Substitution of dextran with PVP led to formation of galactosylated chitosan-graft-PVP (GCPVP)/DNA complexes with negligible cytotoxicity regardless of the concentration of GCPVP and the charge ratio in HepG2 cells.[50] Water-soluble chitosan was modified with a lactobionic acid-bearing galactose group to achieve target specificity and reduce cytotoxicity associated with acetic acid-soluble chitosan.[51] Cytotoxicity studies suggested absence of cytotoxicity with GC-DNA complexes and higher transfection efficiency into HepG2 in contrast to HeLa cells. In another study, low-MW chitosan was modified with lactobionic acid to target liver cells.[52] GC/DNA complex showed highly efficient and cell-selective transfection to hepatocytes, which increased with the improvement of the degree of galactosylation.

PEI was derivatized with terminally galactose-grafted PEG by using a bifunctional PEG derivative containing both an *a-vinyl* sulfone and N-hydroxysuccinimide (NHS) ester groups (VS-PEG-NHS).[53] The transfection efficiency with 1% galactose-PEG-PEI in HepG2 cells was higher than that of PEI; on the other hand, in mouse fibroblasts (NIH3T3), which have no ASGPR, the transfection efficiency drastically decreased to 1/40 of that with PEI. In another study, galactosylated PEI was synthesized with a broad range of substitution, ranging from 3.5 to 31% of PEI amino groups.[54] The transfection efficiency of gal–PEI/DNA complex in HepG2 cells was slightly increased for galactosylated PEIs at a N/P ratio of 2 and strongly reduced at higher N/P ratios. Later, galactosylated PEI was conjugated with PVP to reduce cytotoxicity along with target specificity.[55] The transfection efficiency of the complexes at charge ratio of 40 in the

Published by Woodhead Publishing Limited, 2013

HepG2 was observed to be higher than that of PEI/DNA. Galactosylated PEG-chitosan-graft-PEI complexes were reported to transfect liver cells more effectively than PEI (25 kDa) *in vivo* after intravenous administration.[56]

6.3 Conclusions

Application of target-specific nanoparticles for gene delivery has shown promising results in preclinical studies, demonstrating their potential as therapeutics carriers. Recent progress towards the development of ligand-decorated nanoparticles for target specificity has generated great enthusiasm in both academia and industry. However, there is a long way to go in order to develop formulations for therapeutic use. Nanoparticles with improved intracellular uptake, high target specificity, improved efficacy, and low cytotoxicity profiles in contrast to non-targeted nanoparticles will further pave the way for generation of novel delivery vehicles for therapeutic agents.

6.4 References

1. Owens, D.E., 3rd and Peppas, N.A. (2006) Opsonization, biodistribution, and pharmacokinetics of polymeric nanoparticles. *Int J Pharm*, **307**, 93–102.
2. Panagi, Z., Beletsi, A., Evangelatos, G., Livaniou, E., Ithakissios, D.S., et al. (2001) Effect of dose on the biodistribution and pharmacokinetics of PLGA and PLGA-mPEG nanoparticles. *Int J Pharm*, **221**, 143–52.
3. Durand, R., Paul, M., Rivollet, D., Houin, R., Astier, A., et al. (1997) Activity of pentamidine-loaded methacrylate

Published by Woodhead Publishing Limited, 2013

nanoparticles against Leishmania infantum in a mouse model. *Int J Parasitol*, **27**, 1361–7.

4. Ruehm, S.G., Corot, C., Vogt, P., Kolb, S. and Debatin, J.F. (2001) Magnetic resonance imaging of atherosclerotic plaque with ultrasmall superparamagnetic particles of iron oxide in hyperlipidemic rabbits. *Circulation*, **103**, 415–22.
5. Kim, J.H., Kim, Y.S., Park, K., Lee, S., Nam, H.Y., et al. (2008) Antitumor efficacy of cisplatin-loaded glycol chitosan nanoparticles in tumor-bearing mice. *J Control Release*, **127**, 41–9.
6. Folkman, J. (1995) Angiogenesis in cancer, vascular, rheumatoid and other disease. *Nat Med*, **1**, 27–31.
7. Skinner, S.A., Tutton, P.J. and O'Brien, P.E. (1990) Microvascular architecture of experimental colon tumors in the rat. *Cancer Res*, **50**, 2411–17.
8. Matsumura, Y. and Maeda, H. (1986) A new concept for macromolecular therapeutics in cancer chemotherapy: mechanism of tumoritropic accumulation of proteins and the antitumor agent smancs. *Cancer Res*, **46**, 6387–92.
9. Suzuki, M., Hori, K., Abe, I., Saito, S. and Sato, H. (1981) A new approach to cancer chemotherapy: selective enhancement of tumor blood flow with angiotensin II. *J Natl Cancer Inst*, **67**, 663–9.
10. Maeda, H. and Matsumura, Y. (1989) Tumoritropic and lymphotropic principles of macromolecular drugs. *Crit Rev Ther Drug Carrier Syst*, **6**, 193–210.
11. Iwai, K., Maeda, H. and Konno, T. (1984) Use of oily contrast medium for selective drug targeting to tumor: enhanced therapeutic effect and X-ray image. *Cancer Res*, **44**, 2115–21.
12. Wang, J.L., Tang, G.P., Shen, J., Hu, Q.L., Xu, F.J., et al. (2012) A gene nanocomplex conjugated with monoclonal

Published by Woodhead Publishing Limited, 2013

antibodies for targeted therapy of hepatocellular carcinoma. *Biomaterials*, **33**, 4597–607.

13. Sadeqzadeh, E., Rahbarizadeh, F., Ahmadvand, D., Rasaee, M.J., Parhamifar, L., et al. (2011) Combined MUC1-specific nanobody-tagged PEG-polyethylenimine polyplex targeting and transcriptional targeting of tBid transgene for directed killing of MUC1 over-expressing tumour cells. *J Control Release*, **156**, 85–91.
14. Duan, Y., Yang, C., Zhang, Z., Liu, J., Zheng, J., et al. (2010) Poly(ethylene glycol)-grafted polyethylenimine modified with G250 monoclonal antibody for tumor gene therapy. *Hum Gene Ther*, **21**, 191–8.
15. Uemura, H., Nakagawa, Y., Yoshida, K., Saga, S., Yoshikawa, K., et al. (1999) MN/CA IX/G250 as a potential target for immunotherapy of renal cell carcinomas. *Br J Cancer*, **81**, 741–6.
16. Han, J.Y., Choi, D.S., Kim, C., Joo, H. and Min, C.K. (2008) Selective gene transfer to endometrial cancer cells by a polymer against matrix metalloproteinase 2 (MMP–2). *Cancer Biother Radiopharm*, **23**, 247–58.
17. Germershaus, O., Merdan, T., Bakowsky, U., Behe, M. and Kissel, T. (2006) Trastuzumab-polyethylenimine-polyethylene glycol conjugates for targeting Her2-expressing tumors. *Bioconjug Chem*, **17**, 1190–9.
18. Chiu, S.J., Ueno, N.T. and Lee, R.J. (2004) Tumor-targeted gene delivery via anti-HER2 antibody (trastuzumab, Herceptin) conjugated polyethylenimine. *J Control Release*, **97**, 357–69.
19. Brennan, F., Shaw, L., Wing, M. and Robinson, C. (2004) Preclinical safety testing of biotechnology-derived pharmaceuticals. *Mol Biotechnol*, **27**, 59–74.
20. Weinberg, W., Frazier-Jessen, M., Wu, W., Weir, A., Hartsough, M., et al. (2005) Development and regulation

Published by Woodhead Publishing Limited, 2013

of monoclonal antibody products: Challenges and opportunities. *Cancer Metastasis Rev*, **24**, 569–84.

21. Daniels, T.R., Delgado, T., Rodriguez, J.A., Helguera, G. and Penichet, M.L. (2006) The transferrin receptor part I: Biology and targeting with cytotoxic antibodies for the treatment of cancer. *Clin Immunol*, **121**, 144–58.
22. Ogris, M., Brunner, S., Schuller, S., Kircheis, R. and Wagner, E. (1999) PEGylated DNA/transferrin-PEI complexes: reduced interaction with blood components, extended circulation in blood and potential for systemic gene delivery. *Gene Ther*, **6**, 595–605.
23. Ogris, M., Walker, G., Blessing, T., Kircheis, R., Wolschek, M., et al. (2003) Tumor-targeted gene therapy: strategies for the preparation of ligand-polyethylene glycol-polyethylenimine/DNA complexes. *J Control Release*, **91**, 173–81.
24. Kursa, M., Walker, G.F., Roessler, V., Ogris, M., Roedl, W., et al. (2003) Novel shielded transferrin-polyethylene glycol-polyethylenimine/DNA complexes for systemic tumor-targeted gene transfer. *Bioconjug Chem*, **14**, 222–31.
25. Kircheis, R., Wightman, L., Schreiber, A., Robitza, B., Rossler, V., et al. (2001) Polyethylenimine/DNA complexes shielded by transferrin target gene expression to tumors after systemic application. *Gene Ther*, **8**, 28–40.
26. Kim, E.M., Jeong, H.J., Heo, Y.J., Moon, H.B., Bom, H.S., et al. (2004) Intratumoral injection of 188Re labeled cationic polyethylenimine conjugates: a preliminary report. *J Korean Med Sci*, **19**, 647–51.
27. Kadiyala, I., Loo, Y., Roy, K., Rice, J. and Leong, K.W. (2010) Transport of chitosan-DNA nanoparticles in human intestinal M-cell model versus normal intestinal enterocytes. *Eur J Pharm Sci*, **39**, 103–9.

Published by Woodhead Publishing Limited, 2013

28. Lee, K.M., Lee, Y.B. and Oh, I.J. (2011) Evaluation of PEG-transferrin-PEI nanocomplex as a gene delivery agent. *J Nanosci Nanotechnol*, **11**, 7078–81.
29. Lee, K.M., Kim, I.S., Lee, Y.B., Shin, S.C., Lee, K.C., et al. (2005) Evaluation of transferrin-polyethylenimine conjugate for targeted gene delivery. *Arch Pharm Res*, **28**, 722–9.
30. Li, Y., Ogris, M., Wagner, E., Pelisek, J. and Ruffer, M. (2003) Nanoparticles bearing polyethyleneglycol-coupled transferrin as gene carriers: preparation and in vitro evaluation. *Int J Pharm*, **259**, 93–101.
31. Varner, J.A. and Cheresh, D.A. (1996) Integrins and cancer. *Curr Opin Cell Biol*, **8**, 724–30.
32. Ruoslahti, E. (1996) RGD and other recognition sequences for integrins. *Annu Rev Cell Dev Biol*, **12**, 697–715.
33. Kunath, K., Merdan, T., Hegener, O., Haberlein, H. and Kissel, T. (2003) Integrin targeting using RGD-PEI conjugates for in vitro gene transfer. *J Gene Med*, **5**, 588–99.
34. Sun, Y.X., Zeng, X., Meng, Q.F., Zhang, X.Z., Cheng, S.X., et al. (2008) The influence of RGD addition on the gene transfer characteristics of disulfide-containing polyethyleneimine/DNA complexes. *Biomaterials*, **29**, 4356–65.
35. Tian, H., Lin, L., Chen, J., Chen, X., Park, T.G., et al. (2011) RGD targeting hyaluronic acid coating system for PEI-PBLG polycation gene carriers. *J Control Release*, **155**, 47–53.
36. Suk, J.S., Suh, J., Choy, K., Lai, S.K., Fu, J., et al. (2006) Gene delivery to differentiated neurotypic cells with RGD and HIV Tat peptide functionalized polymeric nanoparticles. *Biomaterials*, **27**, 5143–50.
37. Kleemann, E., Neu, M., Jekel, N., Fink, L., Schmehl, T., et al. (2005) Nano-carriers for DNA delivery to the lung

Published by Woodhead Publishing Limited, 2013

based upon a TAT-derived peptide covalently coupled to PEG-PEI. *J Control Release*, **109**, 299–316.

38. Ma, Z., Wu, X., Cao, M., Pan, W., Zhu, F., et al. (2003) Selection of trkB-binding peptides from a phage-displayed random peptide library. *Sci China C Life Sci*, **46**, 77–86.
39. Garcia-Suarez, O., Hannestad, J., Esteban, I., Sainz, R., Naves, F.J., et al. (1998) Expression of the TrkB neurotrophin receptor by thymic macrophages. *Immunology*, **94**, 235–41.
40. Talvitie, E., Leppiniemi, J., Mikhailov, A., Hytönen, V.P. and Kellomäki, M. (2012) Peptide-functionalized chitosan–DNA nanoparticles for cellular targeting. *Carbohyd Polym*, **89**, 948–54.
41. Prasad, T.K. and Rao, N.M. (2005) The role of plasmid constructs containing the SV40 DNA nuclear-targeting sequence in cationic lipid-mediated DNA delivery. *Cell Mol Biol Lett*, **10**, 203–15.
42. Dean, D.A., Byrd, J.N., Jr. and Dean, B.S. (1999) Nuclear targeting of plasmid DNA in human corneal cells. *Curr Eye Res*, **19**, 66–75.
43. Escriou, V., Carriere, M., Scherman, D. and Wils, P. (2003) NLS bioconjugates for targeting therapeutic genes to the nucleus. *Adv Drug Deliv Rev*, **55**, 295–306.
44. Opanasopit, P., Rojanarata, T., Apirakaramwong, A., Ngawhirunpat, T. and Ruktanonchai, U. (2009) Nuclear localization signal peptides enhance transfection efficiency of chitosan/DNA complexes. *Int J Pharm*, **382**, 291–5.
45. Weitman, S.D., Lark, R.H., Coney, L.R., Fort, D.W., Frasca, V., et al. (1992) Distribution of the folate receptor GP38 in normal and malignant cell lines and tissues. *Cancer Res*, **52**, 3396–401.

46. Morris, V.B. and Sharma, C.P. (2010) Folate mediated histidine derivative of quaternised chitosan as a gene delivery vector. *Int J Pharm*, **389**, 176–85.
47. Shi, Q., Wang, H., Tran, C., Qiu, X., Winnik, F.M., et al. (2011) Hydrodynamic delivery of chitosan-folate-DNA nanoparticles in rats with adjuvant-induced arthritis. *J Biomed Biotechnol*, **2011**, 148763.
48. Park, I.K., Park, Y.H., Shin, B.A., Choi, E.S., Kim, Y.R., et al. (2000) Galactosylated chitosan-graft-dextran as hepatocyte-targeting DNA carrier. *J Control Release*, **69**, 97–108.
49. Park, I.K., Kim, T.H., Park, Y.H., Shin, B.A., Choi, E.S., et al. (2001) Galactosylated chitosan-graft-poly(ethylene glycol) as hepatocyte-targeting DNA carrier. *J Control Release*, **76**, 349–62.
50. Park, I.K., Jiang, H.L., Cook, S.E., Cho, M.H., Kim, S.I., et al. (2004) Galactosylated chitosan (GC)-graft-poly(vinyl pyrrolidone) (PVP) as hepatocyte-targeting DNA carrier: in vitro transfection. *Arch Pharm Res*, **27**, 1284–9.
51. Kim, T.H., Park, I.K., Nah, J.W., Choi, Y.J. and Cho, C.S. (2004) Galactosylated chitosan/DNA nanoparticles prepared using water-soluble chitosan as a gene carrier. *Biomaterials*, **25**, 3783–92.
52. Gao, S., Chen, J., Xu, X., Ding, Z., Yang, Y.H., et al. (2003) Galactosylated low molecular weight chitosan as DNA carrier for hepatocyte-targeting. *Int J Pharm*, **255**, 57–68.
53. Sagara, K. and Kim, S.W. (2002) A new synthesis of galactose-poly(ethylene glycol)-polyethylenimine for gene delivery to hepatocytes. *J Control Release*, **79**, 271–81.
54. Kunath, K., von Harpe, A., Fischer, D. and Kissel, T. (2003) Galactose-PEI-DNA complexes for targeted gene

delivery: degree of substitution affects complex size and transfection efficiency. *J Control Release*, **88**, 159–72.

55. Cook, S.E., Park, I.K., Kim, E.M., Jeong, H.J., Park, T.G., et al. (2005) Galactosylated polyethylenimine-graft-poly(vinyl pyrrolidone) as a hepatocyte-targeting gene carrier. *J Control Release*, **105**, 151–63.
56. Jiang, H.L., Kwon, J.T., Kim, E.M., Kim, Y.K., Arote, R., et al. (2008) Galactosylated poly(ethylene glycol)-chitosan-graft-polyethylenimine as a gene carrier for hepatocyte-targeting. *J Control Release*, **131**, 150–7.

Published by Woodhead Publishing Limited, 2013

7

Polymeric nanoparticles for gene delivery

DOI: 10.1533/9781908818645.137

Abstract: Polymeric nanoparticles are generally engineered via a self-assembly process using block-copolymers which consists of two or more polymer chains varying in hydrophilicity that are usually biocompatible and biodegradable. Nanoparticles being compact are well suited to traverse cellular membranes to mediate gene delivery. It is also expected that due to smaller size, nanoparticles would be less susceptible to RES clearance and will have better penetration into tissues and cells, when used in *in vivo* therapy. However, due to excess cationic charge polymeric nanoparticles electrostatically interact with membrane proteins and induce cytotoxicity. The present chapter provides the pros and cons of nanoparticles mediated gene delivery.

Key words: biocompatible, biodegradable, transgene, polycationic polymers, cytotoxicity, non-biodegradable.

7.1 Introduction

Polymeric nanoparticles are generally derived from biocompatible and biodegradable polymers. These nanoparticles are usually formulated via a self-assembly process using block-copolymers of two or more polymer chains varying in hydrophilicity. The copolymers are rearranged into a core–shell structure in an aqueous environment.[1] The hydrophobic units form the inner core to mask their exposure to the aqueous environment while the hydrophilic units form the outer shell to stabilize the core.[2] This whole rearrangement process results in nanostructures that are well suited for drug/gene delivery. Polymeric nanoparticles can be engineered to entrap either hydrophilic or hydrophobic drug molecules, as well as macromolecules such as proteins and nucleic acids.[3] The properties of polymeric nanoparticles (e.g. size, surface charge, and structure) are dependent on the N/P ratio. Several natural and synthetic polymers such as chitosan, PEI, PLL, PLGA, poly (alkylcyanoacrylate), etc. have been investigated for DNA delivery.

7.2 Advantages of nanoparticles

Due to their small size (usually 10 to 200 nm), polymeric nanoparticles can readily interact with biomolecules on the cell surface or inside the cell. Nanoparticles, being compact, are well suited to traverse cellular membranes to mediate gene delivery. It is also expected that, due to their smaller size, nanoparticles will be less susceptible to RES clearance and will have better penetration into tissues and cells when used in *in vivo* therapy. Also, due to their small size nanoparticles could penetrate tissues such as tumors in depth with a high level of specificity, thereby improving the targeted delivery of drug/gene.[4] Further, nanoparticles prepared from polymers have several advantages, such as ease of manipulation with scope to change the MW, geometry (linear

Published by Woodhead Publishing Limited, 2013

and branched), stability, safety, low cost, and high flexibility regarding the size of transgene delivered. Furthermore, nanoparticles can easily be tagged with various targeting moieties such as RGD peptides or transferrin to achieve target-specific gene delivery.[5,6] Also, various routes of administration are feasible, including oral and by inhalation. These carriers can also be engineered to enable controlled (sustained) drug release from the matrix.

7.3 Limitations of nanoparticles

Although polymeric nanoparticles possess several advantages and have been successfully employed in several studies, they still suffer some limitations. Biocompatibility of polymers employed for nanoparticle preparation was observed to be influenced by different properties of the polymers, such as (i) molecular weight, (ii) charge density and type of cationic functionalities, (iii) structure and sequence (block, random, linear, branched), and (iv) conformational flexibility.[7] The cytotoxicity of polymers was found to be directly correlated to the increase in the MW of the polymer.[7–10] The polycationic polymers constituting nanoparticles undergo strong electrostatic interaction with membrane proteins, which can lead to destabilization and ultimately rupture of the cell membrane.[11] A comparative study between polycationic, neutral, and polyanionic polymers revealed that the polycationic polymers have the highest toxicity, followed by neutral and anionic ones.[11] Also, different types of amine functionalities have been associated with cytotoxicity. The toxic effects of poly-L-glutamic acid derivatives on red blood cells, causing them to agglutinate, have been attributed to the presence of primary amines.[12] Further, macromolecules with tertiary amine groups have been shown to exhibit a

Published by Woodhead Publishing Limited, 2013

lower toxicity than those with primary and secondary residues. Strategies based on reduction of surface charge by coating with HA or PEG have been found to circumvent this problem.[13,14]

Among linear and branched PEIs, the latter are more toxic and less suitable for transfection, particularly at higher N/P ratio.[15] At a N/P ratio of 7.2, toxicity was observed with BPEI (800 and 25 kDa), the mice dying with signs of lung embolism. In contrast, at N/P ratios of 7.2, 6.0, and 4.8, no toxicity was observed with LPEI (22 kDa)/DNA complexes generated in salt or salt-free conditions.[15] Additionally, PEI is a non-biodegradable polymer that can accumulate within the cells, interfering with vital intracellular processes.[16,17] Free PEI can harm cells via growth inhibition or cell death. The complement system can be activated by PEI/DNA complexes if the ratio of cation to anion is high, but the extent of its activation is lowered as the PEI/DNA complexes approach neutrality.[18,19] Chemical modifications by which low-MW PEIs are coupled to generate higher-MW molecules using degradable bi-functional linkages have been found useful in reducing toxicity.

Chitosan has been found to be an efficient *in vitro* and *in vivo* gene delivery vector with no toxicity.[20–24] However, it suffers from low transfection efficiency; optimum conditions for transfection are not clear, and must be well elucidated before its clinical application. Chitosan nanoparticles at a size of 200 nm caused malformations, including a bent spine, pericardial edema, and an opaque yolk in zebrafish embryos.[25] Furthermore, embryos exposed to chitosan nanoparticles showed an increased rate of cell death, high expression of reactive oxygen species, and over-expression of heat shock protein 70, indicating that chitosan nanoparticles can cause physiological stress in zebrafish.[25] Chitosan has been observed to exhibit concentration-dependent cytotoxicity,

Published by Woodhead Publishing Limited, 2013

with half maximal inhibitory concentration (IC_{50}) ranging from 0.2 to 2 mg/ml in most cell models. The IC_{50} values are further dictated by the MW and DDA of chitosan.[26–28] Concerning *in vivo* toxicity, some initial studies done in rabbits and dogs indicated that signs of cytotoxicity began upon subcutaneous injection of doses in the range of 5–50 mg/kg/day.[29,30]

7.4 Conclusions

Though nanomedicine is a relatively new stream of science, its translation into clinical therapeutics has been rapid. This new development of a nanoparticle-based platform holds great promise for treatment of some deadly diseases, such as cancer. Although *in vivo* gene delivery has made significant progress, still many challenges remain to be addressed. There is a need to emphasize strategies to improve transfection efficiency and achieve target-specific delivery. More clinical data are needed to fully substantiate the pros and cons of nanoparticle therapeutics.

7.5 References

1. Zhang, L., Gu, F.X., Chan, J.M., Wang, A.Z., Langer, R.S., et al. (2008) Nanoparticles in medicine: therapeutic applications and developments. *Clin Pharmacol Ther*, 83, 761–9.
2. Farokhzad, O.C., Karp, J.M. and Langer, R. (2006) Nanoparticle-aptamer bioconjugates for cancer targeting. *Expert Opin Drug Deliv*, 3, 311–24.
3. Gu, F., Zhang, L., Teply, B.A., Mann, N., Wang, A., et al. (2008) Precise engineering of targeted nanoparticles

Published by Woodhead Publishing Limited, 2013

by using self-assembled biointegrated block copolymers. *Proc Natl Acad Sci USA*, **105**, 2586–91.

4. Cuenca, A.G., Jiang, H., Hochwald, S.N., Delano, M., Cance, W.G., et al. (2006) Emerging implications of nanotechnology on cancer diagnostics and therapeutics. *Cancer*, **107**, 459–66.
5. Schiffelers, R.M., Ansari, A., Xu, J., Zhou, Q., Tang, Q., et al. (2004) Cancer siRNA therapy by tumor selective delivery with ligand-targeted sterically stabilized nanoparticle. *Nucleic Acids Res*, **32**, e149.
6. Davis, M.E. (2009) The first targeted delivery of siRNA in humans via a self-assembling, cyclodextrin polymer-based nanoparticle: from concept to clinic. *Mol Pharm*, **6**, 659–68.
7. Choksakulnimitr, S., Masuda, S., Tokuda, H., Takakura, Y. and Hashida, M. (1995) In vitro cytotoxicity of macromolecules in different cell culture systems. *J Control Release*, **34**, 233–41.
8. Fischer, D., Bieber, T., Li, Y., Elsasser, H.P. and Kissel, T. (1999) A novel non-viral vector for DNA delivery based on low molecular weight, branched polyethylenimine: effect of molecular weight on transfection efficiency and cytotoxicity. *Pharm Res*, **16**, 1273–9.
9. Morgan, D.M., Larvin, V.L. and Pearson, J.D. (1989) Biochemical characterisation of polycation-induced cytotoxicity to human vascular endothelial cells. *J Cell Sci*, **94 (Pt 3)**, 553–9.
10. Haensler, J. and Szoka, F.C. (1993) Polyamidoamine cascade polymers mediate efficient transfection of cells in culture. *Bioconjug Chem*, **4**, 372–9.
11. Fischer, D., Li, Y., Ahlemeyer, B., Krieglstein, J. and Kissel, T. (2003) In vitro cytotoxicity testing of polycations: influence of polymer structure on cell viability and hemolysis. *Biomaterials*, **24**, 1121–31.

Published by Woodhead Publishing Limited, 2013

12. Dekie, L., Toncheva, V., Dubruel, P., Schacht, E.H., Barrett, L., et al. (2000) Poly-L-glutamic acid derivatives as vectors for gene therapy. *J Control Release*, **65**, 187–202.
13. Jiang, G., Park, K., Kim, J., Kim, K.S., Oh, E.J., et al. (2008) Hyaluronic acid-polyethyleneimine conjugate for target specific intracellular delivery of siRNA. *Biopolymers*, **89**, 635–42.
14. Duan, Y., Yang, C., Zhang, Z., Liu, J., Zheng, J., et al. (2010) Poly(ethylene glycol)-grafted polyethylenimine modified with G250 monoclonal antibody for tumor gene therapy. *Hum Gene Ther*, **21**, 191–8.
15. Wightman, L., Kircheis, R., Rossler, V., Carotta, S., Ruzicka, R., et al. (2001) Different behavior of branched and linear polyethylenimine for gene delivery in vitro and in vivo. *J Gene Med*, **3**, 362–72.
16. Godbey, W.T., Wu, K.K., Hirasaki, G.J. and Mikos, A.G. (1999) Improved packing of poly(ethylenimine)/DNA complexes increases transfection efficiency. *Gene Ther*, **6**, 1380–8.
17. Godbey, W.T., Wu, K.K. and Mikos, A.G. (2001) Poly(ethylenimine)-mediated gene delivery affects endothelial cell function and viability. *Biomaterials*, **22**, 471–80.
18. Ferrari, S., Moro, E., Pettenazzo, A., Behr, J.P., Zacchello, F., et al. (1997) ExGen 500 is an efficient vector for gene delivery to lung epithelial cells in vitro and in vivo. *Gene Ther*, **4**, 1100–6.
19. Plank, C., Mechtler, K., Szoka, F.C., Jr. and Wagner, E. (1996) Activation of the complement system by synthetic DNA complexes: a potential barrier for intravenous gene delivery. *Hum Gene Ther*, **7**, 1437–46.
20. Leong, K.W., Mao, H.Q., Truong-Le, V.L., Roy, K., Walsh, S.M., et al. (1998) DNA-polycation nanospheres

as non-viral gene delivery vehicles. *J Control Release*, **53**, 183–93.

21. Roy, K., Mao, H.Q., Huang, S.K. and Leong, K.W. (1999) Oral gene delivery with chitosan--DNA nanoparticles generates immunologic protection in a murine model of peanut allergy. *Nat Med*, **5**, 387–91.
22. Grenha, A., Grainger, C.I., Dailey, L.A., Seijo, B., Martin, G.P., et al. (2007) Chitosan nanoparticles are compatible with respiratory epithelial cells in vitro. *Eur J Pharm Sci*, **31**, 73–84.
23. Mohammadi, Z., Abolhassani, M., Dorkoosh, F.A., Hosseinkhani, S., Gilani, K., et al. (2011) Preparation and evaluation of chitosan-DNA-FAP-B nanoparticles as a novel non-viral vector for gene delivery to the lung epithelial cells. *Int J Pharm*, **409**, 307–13.
24. Jiang, H., Wu, H., Xu, Y.L., Wang, J.Z. and Zeng, Y. (2011) Preparation of galactosylated chitosan/ tripolyphosphate nanoparticles and application as a gene carrier for targeting SMMC7721 cells. *J Biosci Bioeng*, **111**, 719–24.
25. Hu, Y.L., Qi, W., Han, F., Shao, J.Z. and Gao, J.Q. (2011) Toxicity evaluation of biodegradable chitosan nanoparticles using a zebrafish embryo model. *Int J Nanomedicine*, **6**, 3351–9.
26. Kean, T. and Thanou, M. (2010) Biodegradation, biodistribution and toxicity of chitosan. *Adv Drug Deliv Rev*, **62**, 3–11.
27. Schipper, N.G., Varum, K.M. and Artursson, P. (1996) Chitosans as absorption enhancers for poorly absorbable drugs. 1: Influence of molecular weight and degree of acetylation on drug transport across human intestinal epithelial (Caco–2) cells. *Pharm Res*, **13**, 1686–92.
28. Opanasopit, P., Aumklad, P., Kowapradit, J., Ngawhiranpat, T., Apirakaramwong, A., et al. (2007)

Published by Woodhead Publishing Limited, 2013

Effect of salt forms and molecular weight of chitosans on in vitro permeability enhancement in intestinal epithelial cells (Caco–2). *Pharm Dev Technol*, **12**, 447–55.

29. Carreño-Gómez, B. and Duncan, R. (1997) Evaluation of the biological properties of soluble chitosan and chitosan microspheres. *Int J Pharm*, **148**, 231–40.
30. Minami, S., Oh-oka, M., Okamoto, Y., Miyatake, K., Matsuhashi, A., et al. (1996) Chitosan-inducing hemorrhagic pneumonia in dogs. *Carbohyd Polym*, **29**, 241–6.

Published by Woodhead Publishing Limited, 2013

8

Poly-L-lysine nanoparticles

DOI: 10.1533/9781908818645.147

Abstract: Poly-L-lysine, a cationic polypeptide with amino acid lysine as a repeat unit form electrostatic complexes with DNA, resulting in nanoparticles. The polypeptide chain of PLL possesses an acceptable degree of biodegradability, making it an ideal candidate for *in vivo* gene delivery. PLL appears to be the first polymer developed to substitute viral vectors for gene delivery to avoid the immunogenicity and oncogenicity associated with them. Initial studies observed PLL/DNA complexes as both donut and short stem structures in nanometer size. Several studies have harnessed the immense potential of PLL to deliver genes to *in vitro* and *in vivo* targets. Further, PLL has also been incorporated in different peptides either to introduce cationic amino acids or cell attachment domains.

Key words: poly-L-lysine, polypeptide, asialoorosomucoid, hepatocyte, polystyrene, homobifunctional, amphiphilic.

8.1 Introduction

Since the finding that poly-L-lysine (PLL) forms polyelectrolyte complexes with DNA, it has been widely employed as a

Published by Woodhead Publishing Limited, 2013

non-viral gene delivery vector.[1] It is a cationic polypeptide with amino acid lysine as a repeat unit (Figure 8.1). The degree of polymerization (DP) of lysine can range between 90 and 450 and lead to the formation of a polypeptide chain with an acceptable degree of biodegradability, a property highly desirable for *in vivo* use. As is quite evident from the initial studies, PLL having a MW of less than 3000 Da is unable to form stable complexes with DNA, suggesting that the number of primary amines in the PLL backbone is crucial for complex formation.[2] Though high-MW PLLs resulted in better complexation and yielded efficient gene delivery, the PLL/DNA complexes showed a relatively high cytotoxicity and a tendency to aggregate and precipitate depending on the ionic strength.[3,4] PLL have been used in non-viral vectors as single chain or as oligolysine-containing proteins, thus allowing manipulations in the length of the peptide, sequence, and presentation to enhance transfection efficiency.[5]

Synthesis of low-MW PLL (7500–12 500 Da) was described from N-carboxy-(N^{ε}-benzyloxycarbony1)-L-lysine anhydride (Z-L-lysine NCA) using diethylamine as initiator and DMF as solvent. In another study, Z-L-lysine NCA was polymerized in DMF with triethylamine, diethylamine, or hexylamine as

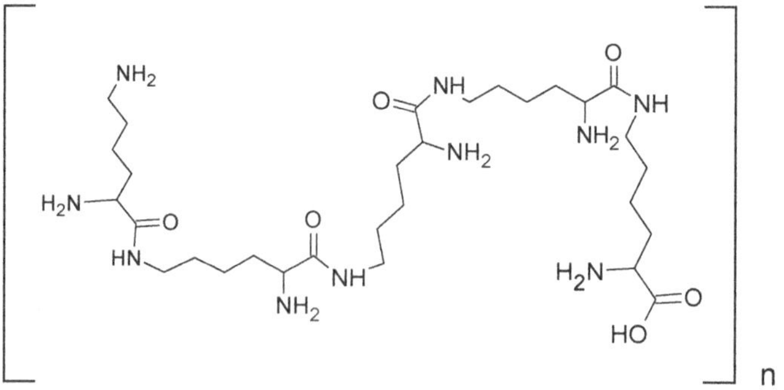

Figure 8.1 Chemical structure of poly-L-lysine

Published by Woodhead Publishing Limited, 2013

initiator, at varying molar ratios of NCA to initiator (M/I ratio) to synthesize PLLs with controlled MW.[6] The degree of polymerization depended linearly on the M/I ratio for both diethylamine and hexylamine, with higher degree of polymerization for the diethylamine-initiated PLL at equal M/I ratio.

8.2 *In vitro* and *in vivo* applications of poly-L-lysine/DNA nanoparticles

One of the seminal studies showed that PLL efficiently condenses DNA and appears as both donuts and short stem structures in electron micrographs, where each of the structures were about three to four layers thick.[1] Wu *et al.* (1987) reported that PLL coupled with asialoorosomucoid formed soluble polyplexes with DNA and efficiently delivered genes to HepG2 cells, which express the receptor for asialoorosomucoid on their surface.[7,8] Further, they described the application of low-MW PLL for gene delivery to hepatocytes and *in vivo* via intravenous administration to mice.[9] Intravenous injection of pDNA complexed to the carrier demonstrated specific hepatic targeting, with 85% of the injected counts being taken up by the liver in 10 min compared with only 17% of the counts when the same amount of uncomplexed radiolabeled DNA was injected under identical conditions. Since PLLs alone exhibit modest to low transfection efficiencies, they have been coupled with endosomal escape agents, such as chloroquine.[10,11] Furthermore, to improve the biological properties of PLL it was grafted with a range of hydrophilic polymer blocks, including PEG, dextran and poly[N-(2-hydroxypropyl) methacrylamide] (pHPMA).[12] The complexes visualized by AFM were observed as discrete particles, typically ~100 nm

Published by Woodhead Publishing Limited, 2013

in diameter. Shielding of the surface by the presence of the hydrophilic polymer resulted in decreased cytotoxicity as compared with simple PLL/DNA complexes, along with improved transfection efficiency.[12] In another study, Choi et al. synthesized PEG-grafted PLL (MW 52.5 kDa) with three different PEG-grafting ratios (5, 10 and 25 mole %).[13] These comb-shaped PEG-g-PLL copolymers showed a 5–30-fold increase in transfection efficiency compared with PLL alone in HepG2 cells.

To develop a hepatocyte-specific gene vector, galactose was introduced into PLL to target ASGPR present on the surface of liver cells.[14] After intravenous administration, [^{32}P] plasmid CAT/Gal-PLL complexes were rapidly cleared from the circulation and preferentially taken up by the liver's parenchymal cells. For efficient gene delivery into cells, PEG-grafted PLL was linked with KALA, which is capable of disrupting endosomal membrane.[15] DNA/PEG-g-PLL complexes remained unaltered at a size of ~200 nm, even with an increase in the KALA amount. Enhanced transfection efficiency was observed with increasing amount of KALA incorporation into PEG-g-PLL/DNA complexes.[15] Further, to enhance intracellular delivery of DNA, a linear reducible polycation based on the oxidative polycondensation of Cys-Lys$_{10}$-Cys peptide was developed.[16] Reducible polycation not only reversibly condensed DNA but also released DNA into the intracellular environment by polyplex-controlled cleavage. A 187-fold higher gene expression level indicated that intracellular delivery of DNA was more efficient using reducible polycation/DOTAP compared with vectors based on non-reducible PLL.[16]

Polystyrene nanoparticles coated with PLL were employed to enhance the efficacy of a DNA vaccine by using the sOVA-C1 plasmid, which contains the gene for chicken egg ovalbumin under the CMV promoter.[17] Intradermal

Published by Woodhead Publishing Limited, 2013

administration of DNA complexed with PLL-coated polystyrene nanoparticles induced high levels of CD8 T-cells as well as ovalbumin-specific antibodies in C57BL/6 mice and inhibited tumor growth after challenge with the ovalbumin-expressing EG7 tumor cell line. Moreover, vaccine efficacy was observed to be dependent on the size of the particles used as well as on the presence of the PLL linker. It was evidenced that PLL-coated polystyrene nanoparticles of size 50 nm were highly effective for the delivery of DNA vaccines as compared with 100 and 200-nm nanoparticles.[17] In another study, a naturally occurring lipid, palmitic acid (PA), was coupled to PLL (PLL-PA) to enhance gene transfer efficacy in bone marrow stromal cells (BMSC).[18] PLL-PA formed condensed structures with pEGFP of size below 700 nm and effectively delivered the pDNA into the nucleus within 5 h of incubation with the cells. PLL-PA delivered the pEGFP to ~80% of the cells, resulting in a maximum transfection efficiency of ~22%. Moreover, two to three times additional dosing during the 24 h period increased the transfection efficiency by two to threefold, without compromising cell viability.[18]

To promote brain targeting, a 30-amino acid peptide, leptin30, derived from an endogenic hormone was covalently conjugated to dendrigraft poly-L-lysine (DGL) via a homobifunctional PEG linker.[19] Complexation of DGL-PEG-Leptin30 with plasmid DNA resulted in nanoparticles of size below 150 nm. The cellular uptake efficiency increased with the degree of PEGylation and brain-targeting ligand conjugation to the DGL-PEG-Leptin30/DNA nanoparticles from none to median. Nanoparticles could be transported across an *in vitro* blood brain barrier (BBB) model, which consists of a monolayer of brain capillary endothelial cells (BCECs), more effectively than DGL-PEG/DNA and DGL/DNA nanoparticles due to significantly higher permeability

Published by Woodhead Publishing Limited, 2013

of DGL-PEG-Leptin30/DNA nanoparticles as compared with other nanoparticles.[19] Further, transfection studies in brain parenchyma microglia cells such as BV–2 cells expressing leptin receptors revealed that the transfection efficacy of DGL-PEG-Leptin30/DNA nanoparticles was higher than that of DGL/DNA and DGL-PEG/DNA nanoparticles and comparable to that of Lipofectamine 2000. Additionally, *in vivo* administration of nanoparticles resulted in reporter gene expression visible in different brain regions (cortical layer, hippocampus, caudate putamen, and substantia nigra), with a preference for DGL-PEG-Leptin30/DNA nanoparticles in the brain when compared with the other two controls.[19]

To blend the advantages of PLL, i.e. its good DNA-binding ability, and chitosan, i.e. its good biocompatibility, copolymers with different PLL grafting ratios (Chi-g-PLL) were synthesized.[20] The Chi-g-PLL polyplexes were observed to be in the size range of 120–200 nm, which was comparable to that of PLL polyplexes and smaller than that of chitosan ones at certain polymer to pDNA weight ratios. A positive correlation between the transfection efficiency of the copolymers and the PLL graft length was found.[20] *In vitro* transfection experiments in 293T cells showed that, with increase in PLL chain length from 2.0 to 5.7, the EGFP-positive cell percentage was improved by a factor of 3.5 (from 6.8 to 23.5%). Furthermore, in HeLa cells, a 3.3-fold increase was reported when the PLL chain length was increased from 2.0 to 5.7, and further increasing the PLL chain length to 15.6 led to a 1.8-fold increase in EGFP-positive cell percentage. However, under optimal conditions, the copolymer Chi-g-$PLL_{32.2}$ transfected 71% of HeLa cells and 56% of 293T cells, respectively. Also, chitosan and Chi-g-PLL polyplexes resulted in cell viabilities over 80% with polymer to pDNA weight ratios varying from 20 to 5.[20]

Published by Woodhead Publishing Limited, 2013

To identify possible effects of polymer MW and architecture on both immediate and delayed cytotoxicity and to provide mechanistic details, *in vitro* cytotoxicity of a library of three structural polylysine variants, namely, linear polylysine (LPL), dendritic polylysine (DPL), and hyperbranched polylysine (HBPL), was evaluated in Chinese hamster ovary (CHO) DG44 cells.[21] At similar MW, the EC_{50} values for the LPL analogues were ~5–250 times higher as compared with the DPL and HBPL samples. For low-MW polycations, osmotic shock was found to be an important contributor to immediate cell death, whereas for the higher-MW analogues direct cell membrane disruption was identified to play a role. Apoptosis was found to be more pronounced for DPL and HBPL as compared with LPL at comparable MWs. This difference was due to the fact that LPL is completely degradable enzymatically, in contrast to DPL and HBPL, which also contain ε-peptidic bonds and are only partially degradable.[21]

A unique mPEG-SS-PLL_{15}-star catiomer was engineered as dual stimulus–responsive, which results in redox-sensitive removal of an external PEG shell induced by acid from the endosomal compartment.[22] The PEG shell removal enhanced the intracellular uptake of mPEG-SS-PLL_{15}-star/DNA complexes in the presence of tumor-relevant glutathione (GSH) concentration, whereas the acid-induced cleavage was to accelerate the release of genetic payload following successful internalization into targeted cells. The *in vitro* cytotoxicity of fabricated catiomers evaluated in 293T and HeLa ceil lines revealed cell viabilities of more than 90% following 4 h incubation at all the concentrations used.[22] The particle size distribution studies showed near-Gaussian distribution of polyplexes between 100 and 350 nm, with a mean diameter of ~200 nm. Finally, *in vitro* transfection efficiency using luciferase and GFP demonstrated comparable results to those obtained with BPEI (25 kDa).[22]

Published by Woodhead Publishing Limited, 2013

To explore a nucleolin-independent pathway for transfection, pH-responsive DNA nanoparticles were formulated by inserting poly-L-histidine between PEG and PLL to form a triblock copolymer system, PEG-$CH_{12}K_{18}$.[23] Insertion of poly-L-histidine increased the buffering capacity of PEG-$CH_{12}K_{18}$ to levels comparable with BPEI. DNA was condensed into rod-shaped nanoparticles by PEG-$CH_{12}K_{18}$ and had similar morphology and colloidal stability as PEG-PLL copolymer (PEG-CK_{30})/DNA nanoparticles. *In vitro* uptake studies suggested that PEG-$CH_{12}K_{18}$-DNA nanoparticles internalized human bronchial epithelial cells (BEAS–2B) that lack surface nucleolin by a clathrin-dependent endocytic mechanism followed by endo-lysosomal processing.[23] Though PEG-$CH_{12}K_{18}$-DNA nanoparticles followed a degradative endo-lysosomal pathway, ~20-fold improvement in *in vitro* transfection efficiency was observed in contrast to (PEG-CK_{30})-DNA nanoparticles. Further, *in vivo* gene delivery to lung airways in BALB/c mice was improved by approximately threefold, while maintaining a low toxicity.[23]

Polymer replica particles have been explored for the adsorption and co-adsorption of therapeutics for their concurrent delivery. Replica particles based on PLL (PLL_{RP}) polymers cross-linked via a homobifunctional linker were designed to support co-adsorption of a pDNA (SPT7pTL, a vector expressing the human SPT7 nuclear transcription factor) and a peptide hormone (α-melanocyte-stimulating hormone (α-MSH)) for concurrent transfection and induction of a cellular function.[24] The size of the PLL_{RP} particles, as analyzed by TEM, was ~1.5 μm. The efficacy of PLL_{RP} at delivering SPT7pTL DNA into B16-F1 melanoma cells and melanin stimulation was investigated. The PLL_{RP}/SPT7pTL complex showed a significant increase (~60–70%) in gene expression of SPT7 in terms of cell number, and the melanin

Published by Woodhead Publishing Limited, 2013

production was 20-fold higher than in the controls. Further, the cell viability was found to be > 90% for the complexes.[24]

To engineer an advanced gene delivery system retaining the advantages of both PEI and PLL while avoiding their shortcomings, a ternary copolymer of PEG-b-PLL-g-LPEI (PPI) was synthesized with different LPEI graft densities.[25] Further, to employ PPI for target-specific gene delivery, folate was attached to it. The polyplexes prepared within the N/P range of 8–25 based on PPI2 (graft density 6.3) and PPI3 (graft density 5.9) displayed positive charge lower than +5 mV and particle size less than 200 nm. A transfection study carried out in three cell lines (HepG2, U251, and BHK21) using a reporter gene assay strongly demonstrated that grafting a proper amount of short linear PEI chains to PEGylated PLL can remarkably improve the vector's gene transfection efficiency. PPI3 with 57% linear PEI displayed the highest transfection efficacy among the three synthesized PPIs.[25]

Gene delivery efficiency of PLL was improved by imparting an amphiphilic property to PLL by substituting ~10% of ε-NH_2 with several endogenous lipids of variable chain lengths (lipid carbon chain ranging from 8 to 18).[26] High-MW (~25 vs. 4 kDa) lipid-modified PLL was found to be more effective in delivering plasmid DNA intracellularly in clinically relevant bone marrow stromal cells. In the case of lipid-substituted 25 kDa PLL, the efficacy of pDNA delivery was seen to be dependent on the extent of lipid substitution. Myristic, palmitic, and stearic acid-substituted polymers resulted in highly efficient DNA delivery; the effect was ascribed to high substitution ratios obtained with these lipids (~10 lipids/PLL). Transfection data revealed that amphiphilic PLLs significantly improved (20–25%) the transfection efficiency in comparison to native PLL and the commercial transfection agent Lipofectamine 2000. Also, the transfection

Published by Woodhead Publishing Limited, 2013

efficiency of the polymers was dependent on the extent of lipid substitution.[26]

To highlight the effect of DNA vector topology when complexed to PLL and its quantification in transfection efficiency, cell uptake followed by transfection efficiency studies of PLL/DNA complexes were done in CHO cells.[27] Complexation of supercoiled pDNA with PLL resulted in a polyplex with a mean diameter of 139.06 nm, while open circular and linear pDNA counterparts yielded mean diameters of 305.54 and 841.5 nm, respectively. Moreover, complexes comprising supercoiled pDNA were more resistant to nuclease degradation than other topological counterparts. Uptake data suggests that 1 h post transfection 61% of supercoiled-pDNA polyplexes were associated with the nucleus, in comparison to open circular (24.3%) and linear-pDNA polyplexes (3.5%), respectively.[27] Furthermore, supercoiled-pDNA polyplexes showed highest transfection efficiency of 41%, in contrast to 18.6% for linear-pDNA polyplexes.

8.3 Polylysine-containing peptides for gene delivery

In order to incorporate cationic amino acids or cell attachment domains, PLLs can be inserted into other peptides or proteins, thereby allowing a fine control of synthesis and toxicity reduction.[28–30] To test the influence of lysine content on DNA condensation and transfection efficiency, cationic peptides possessing a single cysteine, tryptophan, and lysine repeat were synthesized.[31] The N-terminal cysteine in each peptide was either alkylated or oxidatively dimerized to produce peptides possessing lysine chains of 3–36 residues. It was demonstrated that peptides with lysine repeats of 13 or

Published by Woodhead Publishing Limited, 2013

more effectively condensed DNA and resulted in formation of nanoparticles in the size range of 53–230 nm, while shorter peptides containing eight or fewer lysine residues formed large complexes ranging from 0.7 to 3 μm.[31] Further, a 1000-fold increase in gene expression was observed in HepG2 cells when transfected with DNA condensates prepared with alkylated Cys-Trp-Lys18 (AlkCWK18) versus polylysine19.[31] In another study, McKenzie et al. compared a batch of lysine-containing peptides, namely $AlkCWK_{18}$, $AlkCYK_{18}$ and K_{20}.[32] The study was designed to test the possible effect of insertion of the aromatic amino acids tyrosine and tryptophan with respect to DNA-binding affinity. It was deduced that aromatic amino acid substitution did not have a vital role in DNA binding, condensation, and gene transfer efficacy. Furthermore, four different variants of CWK_{20} peptide differing in the number of cysteine residues (one to four) inserted into the peptide were designed.[32] Upon binding with DNA, cysteine residues oxidized spontaneously, and inter-peptide disulfide bonds prevented DNA from dissociating, leading to smaller nanoparticles below 50 nm. After internalization, the reducing cell cytoplasm milieu favored the relaxation of the polyplex, allowing rapid release of DNA. The gene expression level obtained with $CWK_{17}C$ was found to be 60-fold higher than that reached by $AlkCWK_{18}$.[32]

In order to establish the minimum number of lysines required to efficiently condense DNA and achieve high levels of gene expression, lysine-based peptides with sequence YKAKnWK ($n = 4–12$) were engineered.[29] It was demonstrated that peptide sequence $YKAK_8K$, which contained ten lysines and a tryptophan, resulted in the best variant in a variety of cell lines when assembled with a membrane-destabilizing peptide. Further, to explore the potential of synthetic peptides as DNA-binding and DNA-compacting agents for receptor-mediated gene delivery, a series of branched oligocationic

Published by Woodhead Publishing Limited, 2013

peptides that differed in the number and type (lysine, arginine, ornithine) of cationic amino acids were synthesized.[5] The minimal chain length for the formation of DNA complexes capable of receptor-mediated gene delivery was found to be of six lysine residues. The signature characteristic of these branched peptides includes a terminal glycine acting as an attachment point for effectors, ligands, and stabilizers like PEG at the C-terminus.[5]

8.4 Conclusions

PLL was one of the first polymers developed to substitute viral vectors for gene delivery to avoid the immunogenicity and oncogenicity of the latter. Peptide–DNA interactions occur by virtue of several different mechanisms, such as hydrogen bond formation, hydrophobic or electrostatic interactions, and water extrusion effects. Although significant progress has been made in gene transfer techniques using PLL during the last two decades, the prerequisites for clinical use in terms of efficiency and specificity have not yet been met. Studies to bypass sub-cellular barriers, such as endosomal escape and nuclear translocation, need to be conducted while designing PLL-based nanoparticles for gene delivery.

8.5 References

1. Laemmli, U.K. (1975) Characterization of DNA condensates induced by poly(ethylene oxide) and polylysine. *Proc Natl Acad Sci U S A*, 72, 4288–92.
2. Kwoh, D.Y., Coffin, C.C., Lollo, C.P., Jovenal, J., Banaszczyk, M.G., et al. (1999) Stabilization of poly-L-

Published by Woodhead Publishing Limited, 2013

lysine/DNA polyplexes for in vivo gene delivery to the liver. *Biochim Biophys Acta*, **1444**, 171–90.

3. Choi, Y.H., Liu, F., Kim, J.S., Choi, Y.K., Park, J.S., et al. (1998) Polyethylene glycol-grafted poly-L-lysine as polymeric gene carrier. *J Control Release*, **54**, 39–48.
4. Liu, G., Molas, M., Grossmann, G.A., Pasumarthy, M., Perales, J.C., et al. (2001) Biological properties of poly-L-lysine-DNA complexes generated by cooperative binding of the polycation. *J Biol Chem*, **276**, 34379–87.
5. Plank, C., Tang, M.X., Wolfe, A.R. and Szoka, F.C., Jr. (1999) Branched cationic peptides for gene delivery: role of type and number of cationic residues in formation and in vitro activity of DNA polyplexes. *Hum Gene Ther*, **10**, 319–32.
6. Van Dijk-Wolthuis, W.N.E., van de Water, L., van de Wetering, P., Van Steenbergen, M.J., Kettenes-van den Bosch, J.J., et al. (1997) Synthesis and characterization of poly-L-lysine with controlled low molecular weight. *Macromol Chem Physic*, **198**, 3893–906.
7. Wu, G.Y. and Wu, C.H. (1988) Evidence for targeted gene delivery to Hep G2 hepatoma cells in vitro. *Biochemistry*, **27**, 887–92.
8. Wu, G.Y. and Wu, C.H. (1987) Receptor-mediated in vitro gene transformation by a soluble DNA carrier system. *J Biol Chem*, **262**, 4429–32.
9. Wu, G.Y. and Wu, C.H. (1988) Receptor-mediated gene delivery and expression in vivo. *J Biol Chem*, **263**, 14621–4.
10. Martin, M.E. and Rice, K.G. (2007) Peptide-guided gene delivery. *AAPS J*, **9**, E18–29.
11. Wolfert, M.A. and Seymour, L.W. (1998) Chloroquine and amphipathic peptide helices show synergistic transfection in vitro. *Gene Ther*, **5**, 409–14.

Published by Woodhead Publishing Limited, 2013

12. Toncheva, V., Wolfert, M.A., Dash, P.R., Oupicky, D., Ulbrich, K., et al. (1998) Novel vectors for gene delivery formed by self-assembly of DNA with poly(L-lysine) grafted with hydrophilic polymers. *Biochim Biophys Acta*, **1380**, 354–68.
13. Choi, Y.H., Liu, F., Kim, J.S., Choi, Y.K., Park, J.S., et al. (1998) Polyethylene glycol-grafted poly-L-lysine as polymeric gene carrier. *J Control Release*, **54**, 39–48.
14. Hashida, M., Takemura, S., Nishikawa, M. and Takakura, Y. (1998) Targeted delivery of plasmid DNA complexed with galactosylated poly(L-lysine). *J Control Release*, **53**, 301–10.
15. Lee, H., Jeong, J.H. and Park, T.G. (2002) PEG grafted polylysine with fusogenic peptide for gene delivery: high transfection efficiency with low cytotoxicity. *J Control Release*, **79**, 283–91.
16. Read, M.L., Bremner, K.H., Oupicky, D., Green, N.K., Searle, P.F., et al. (2003) Vectors based on reducible polycations facilitate intracellular release of nucleic acids. *J Gene Med*, **5**, 232–45.
17. Minigo, G., Scholzen, A., Tang, C.K., Hanley, J.C., Kalkanidis, M., et al. (2007) Poly-L-lysine-coated nanoparticles: a potent delivery system to enhance DNA vaccine efficacy. *Vaccine*, **25**, 1316–27.
18. Clements, B.A., Incani, V., Kucharski, C., Lavasanifar, A., Ritchie, B., et al. (2007) A comparative evaluation of poly-L-lysine-palmitic acid and Lipofectamine 2000 for plasmid delivery to bone marrow stromal cells. *Biomaterials*, **28**, 4693–704.
19. Liu, Y., Li, J., Shao, K., Huang, R., Ye, L., et al. (2010) A leptin derived 30-amino-acid peptide modified pegylated poly-L-lysine dendrigraft for brain targeted gene delivery. *Biomaterials*, **31**, 5246–57.

Published by Woodhead Publishing Limited, 2013

20. Yu, H., Deng, C., Tian, H., Lu, T., Chen, X., et al. (2011) Chemo-physical and biological evaluation of poly(L-lysine)-grafted chitosan copolymers used for highly efficient gene delivery. *Macromol Biosci*, **11**, 352–61.
21. Kadlecova, Z., Baldi, L., Hacker, D., Wurm, F.M. and Klok, H.A. (2012) Comparative study on the in vitro cytotoxicity of linear, dendritic, and hyperbranched polylysine analogues. *Biomacromolecules*, **13**, 3127–37.
22. Cai, X., Dong, C., Dong, H., Wang, G., Pauletti, G.M., et al. (2012) Effective gene delivery using stimulus-responsive catiomer designed with redox-sensitive disulfide and acid-labile imine linkers. *Biomacromolecules*, **13**, 1024–34.
23. Boylan, N.J., Kim, A.J., Suk, J.S., Adstamongkonkul, P., Simons, B.W., et al. (2012) Enhancement of airway gene transfer by DNA nanoparticles using a pH-responsive block copolymer of polyethylene glycol and poly-l-lysine. *Biomaterials*, **33**, 2361–71.
24. Zhang, X., Oulad-Abdelghani, M., Zelkin, A.N., Wang, Y., Haikel, Y., et al. (2010) Poly(L-lysine) nanostructured particles for gene delivery and hormone stimulation. *Biomaterials*, **31**, 1699–706.
25. Dai, J., Zou, S., Pei, Y., Cheng, D., Ai, H., et al. (2011) Polyethylenimine-grafted copolymer of poly(l-lysine) and poly(ethylene glycol) for gene delivery. *Biomaterials*, **32**, 1694–705.
26. Incani, V., Lin, X., Lavasanifar, A. and Uludag, H. (2009) Relationship between the extent of lipid substitution on poly(L-lysine) and the DNA delivery efficiency. *ACS Appl Mater Interfaces*, **1**, 841–8.
27. Dhanoya, A., Chain, B.M. and Keshavarz-Moore, E. (2011) The impact of DNA topology on polyplex uptake and transfection efficiency in mammalian cells. *J Biotechnol*, **155**, 377–86.

Published by Woodhead Publishing Limited, 2013

28. Haines, A.M., Irvine, A.S., Mountain, A., Charlesworth, J., Farrow, N.A., et al. (2001) CL22 – a novel cationic peptide for efficient transfection of mammalian cells. *Gene Ther*, **8**, 99–110.
29. Gottschalk, S., Sparrow, J.T., Hauer, J., Mims, M.P., Leland, F.E., et al. (1996) A novel DNA-peptide complex for efficient gene transfer and expression in mammalian cells. *Gene Ther*, **3**, 448–57.
30. McKenzie, D.L., Kwok, K.Y. and Rice, K.G. (2000) A potent new class of reductively activated peptide gene delivery agents. *J Biol Chem*, **275**, 9970–7.
31. Wadhwa, M.S., Collard, W.T., Adami, R.C., McKenzie, D.L. and Rice, K.G. (1997) Peptide-mediated gene delivery: influence of peptide structure on gene expression. *Bioconjug Chem*, **8**, 81–8.
32. McKenzie, D.L., Collard, W.T. and Rice, K.G. (1999) Comparative gene transfer efficiency of low molecular weight polylysine DNA-condensing peptides. *J Pept Res*, **54**, 311–18.

Published by Woodhead Publishing Limited, 2013

9

Chitosan nanoparticles

DOI: 10.1533/9781908818645.163

Abstract: Chitosan an aminoglucopyran, is composed of randomly distributed *N*-acetylglucosamine and β-(1,4)-linked glucosamine residues. The transfection efficiency of chitosan/DNA nanoparticles depends on several factors such as the degree of deacetylation and molecular weight of the chitosan, pH, protein interactions, charge ratio of chitosan to DNA (N/P ratio), cell type, nanoparticle size and interactions with cells. The DNA binding affinity and transfection efficiency have been found to increase with increase in DDA or MW while maximum protein expression levels are achieved by obtaining an intermediate stability through control of MW and DDA. So far, chitosan has been implicated in several *in vitro* and *in vivo* gene delivery applications. This chapter provides detailed account of applications of chitosan in gene delivery.

Key words: chitosan, glucosamine, degree of deacetylation, complexes, N/P ratio, chitosan salt, serum.

9.1 Introduction

Chitosan-based vectors have emerged as one of the potential non-viral vectors that can safely deliver genetic materials

Published by Woodhead Publishing Limited, 2013

including pDNA, ODNs and siRNA. Some of the advantages associated with chitosan are low toxicity, low immunogenicity, and high biocompatibility along with a high cationic charge.[1,2] Due to its positive charge, chitosan can readily form polyelectrolyte complexes with negatively charged nucleotides by electrostatic interaction. However, its clinical application is hampered due to low specificity and transfection efficiency. Further, the formulation parameters significantly affect the gene delivery efficacy of chitosan.

Chitosan, an aminoglucopyran, is composed of randomly distributed *N*-acetylglucosamine and β-(1,4)-linked glucosamine residues. To explore the vast spectrum of applicability of this polysaccharide, enormous attempts have been made to functionalize and derivatize it. Chitin is the second most abundant natural biopolymer, next only to cellulose, and is isolated from exoskeletons of crustaceans (crabs, shrimps, etc.), cell walls of fungi, and insects. Chemically, chitin is a linear cationic heteropolymer of randomly distributed *N*-acetylglucosamine and glucosamine residues with β–1,4 linkage. Chitosan is produced by *N*-deacetylation of chitin in the presence of alkali (Figure 9.1). Controlled derivatization of chitin results in chitosan with DDA between 40% to 98% and MW between 5×10^4 Da and 2×10^6 Da.[3] The biological application of chitosan depends on DDA and DP, which also decide the MW of the polymer. Chitosan possess reactive hydroxyl and amino groups and is usually less crystalline than chitin. On heating, it degrades prior to melting; thus these polymers have no melting point. It can be considered as a strong base, as it possesses primary amino groups with a pKa value of 6.3. Due to the presence of amino groups, the charged state and properties of chitosan are dictated by the pH.[4] At low pH, the amino groups are protonated and become positively charged, making chitosan a water-soluble cationic

Published by Woodhead Publishing Limited, 2013

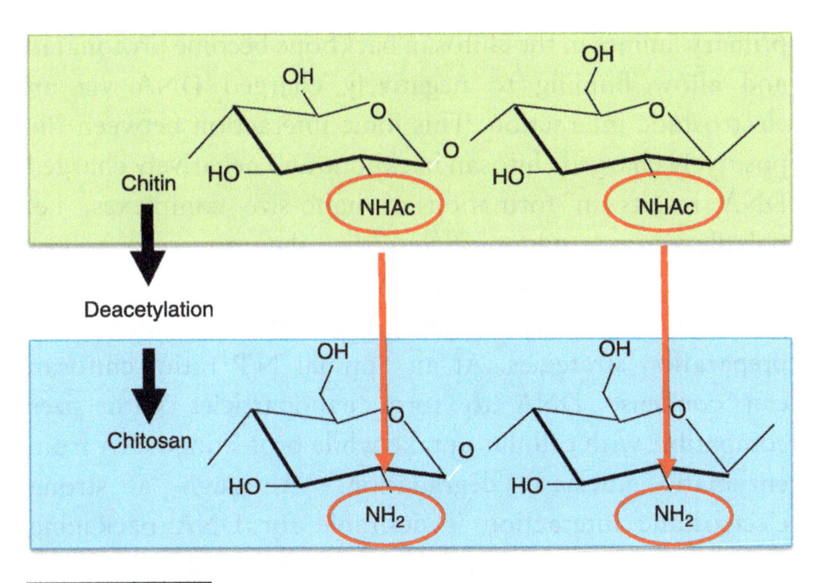

Figure 9.1 Preparation of chitosan from chitin

polyelectrolyte. On the other hand, as the pH is increased above 6, chitosan amino groups become deprotonated; the polymer loses its charge and becomes insoluble. This transition between solubility and insolubility occurs at its pKa value, between pH 6 and 6.5. Hence, chitosan is readily soluble in acidic media such as acetic acid, citric acid, glutamic acid, aspartic acid, hydrochloric acid, and lactic acid, and insoluble at neutral and alkaline pH values. Along with pH, the solubility of chitosan is also dictated by the DDA, MW and ionic strength of the solution. Under physiological conditions, chitosan is digested either by lysozymes or by chitinases, which can be produced by the normal flora in the human intestine or exist in the blood.[5–7] Due to these properties, it has been widely employed for drug/gene delivery both in pharmaceutical research and in industry.[8]

The gene delivery aspect of chitosan is due to the presence of a cationic charge. At acidic pH, below the pKa, the

Published by Woodhead Publishing Limited, 2013

primary amines in the chitosan backbone become protonated and allow binding to negatively charged DNA via an electrostatic interaction. This ionic interaction between the positively charged chitosan backbone and negatively charged DNA results in formation of nano-size complexes, i.e. polyplexes or nanoparticles, in the aqueous milieu (Figure 9.2). Furthermore, DNA can also be entrapped into the chitosan matrix by using conventional nanoparticle preparation strategies. At an optimal N/P ratio, chitosan can condense DNA to form nanoparticles with sizes compatible with cellular uptake while protecting DNA from enzymatic nuclease degradation.[9] Although a strong electrostatic interaction is desirable for DNA packaging and protection, it may pose a challenge for DNA release once the nanoparticles reach the site of action; hence a subtle

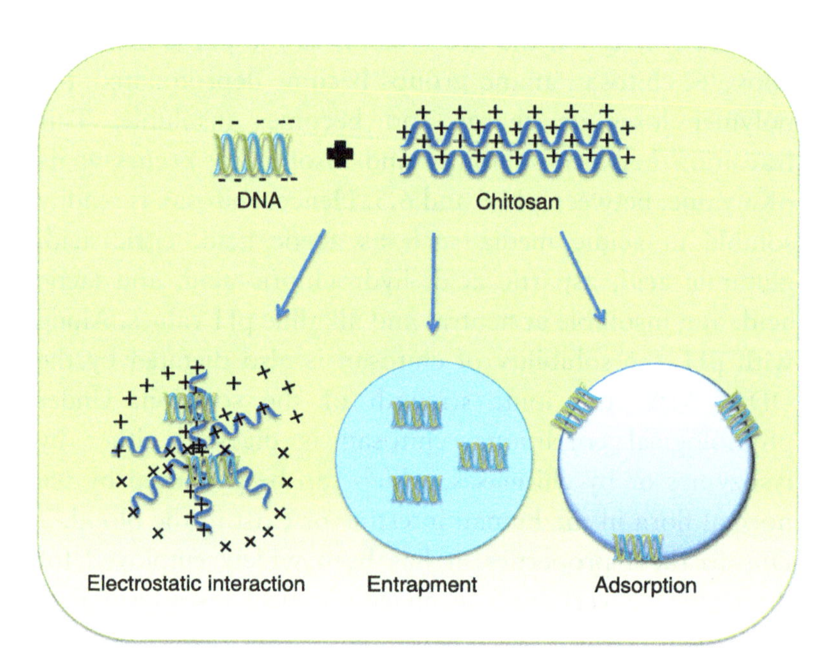

Figure 9.2 Different types of chitosan nanoparticles prepared from interaction of chitosan with DNA

Published by Woodhead Publishing Limited, 2013

balance can be achieved by manipulating the formulation-related parameters.

9.2 Factors affecting transfection efficiency of chitosan nanoparticles

The immense potential of chitosan for *in vitro* gene delivery was first reported by Mumper et al.[10] Chitosan nanoparticles can be prepared by several different methods, such as coacervation, ionic gelation, covalent cross-linking, and desolvation.[11–14] The variability in preparation methodology, together with other parameters like pH, charge ratios, DNA concentrations, salt concentrations, and coacervation temperatures, can significantly affect the size of the resultant chitosan/DNA nanoparticles, and hence the transfection efficiency.[15–20] Hence, considerable attention should be paid to these factors while designing a gene delivery system.

Although chitosan has been widely employed for DNA delivery, it has also been reported to enhance the transfectivity of other transfecting agents. Low MW chitosan was observed to act as a condensing agent and enhance transfection efficacy twofold in HepG2 cells, mediated by a cationic emulsion composed of the following reagents: 3β [N-(N',N'-dimethylaminoethane) carbamoyl] cholesterol, Tween 80, dioleoylphosphatidylethanolamine, and castor oil.[2] Chitosan adsorbed onto lactose-lipid:polycation:pDNA particles by spray-drying was shown to impart improved transfection efficacy and *in vitro* deposition on human lung epithelial carcinoma, A549 cells.[21] Furthermore, chitosan in complexation with adenovirus was found to facilitate adenoviral transfection in mammalian cells lacking viral receptors.[22]

Published by Woodhead Publishing Limited, 2013

9.2.1 Molecular weight

The MW of chitosan is an important parameter that dictates the chitosan/DNA nanoparticles' stability, particle size, cellular uptake efficiency, release of DNA from nanoparticles after endocytosis, and hence the transfection efficacy of the nanoparticles.[23,24] A decrease in the size of nanoparticles has been observed with a decrease in the MW of the chitosan.[23]

Another study reported that the mean particle size of chitosan nanoparticles decreased from 181 to 155 nm when the MW dropped from 213 to 48 kDa.[9] However, as the MW decreased further a reversal in the size trend was observed, i.e. chitosans of 17 and 10 kDa resulted in nanoparticles with an average size of 269 and 289 nm, respectively. It could be inferred from these studies that chitosan of an optimal MW should be chosen to attain the desirable particle size, since the transfection efficacy of the nanoparticles strongly depends on the size of the particles.[25] Cellular uptake has been well correlated with gene expression for chitosans of 40, 84 and 100 kDa complexed with the pGL3-Luc plasmid in SOJ cells.[11]

Efficient transfection was observed with chitosans of MW 20, 45 and 200 kDa in CHO-K1 cells, while higher MW chitosan (460 kDa) showed only slightly higher transfection efficiency than naked DNA due to the large particle size, which resulted in decreased cellular uptake.[12] Also, chitosans of MW 15 and 52 kDa promoted higher luciferase expression in A549, B16 melanoma, and HeLa cells.[24] On the other hand, chitosan heptamer (1.3 kDa) did not show any transfection efficiency, probably due to particle aggregation. Furthermore, transfection efficiency observed with chitosan of 100 kDa was significantly lower than for chitosan of 15 and 52 kDa in all cell lines, despite comparable particle size.[24] Also, chitosan oligomer (18 monomer units) was shown to have higher transfection efficiency than high MW

Published by Woodhead Publishing Limited, 2013

chitosan (162 kDa).[13] Additionally, low MW chitosan (10 kDa) was also shown to yield higher transfection efficiency than the higher MW chitosans (40, 80, 150 kDa) when employed to deliver luciferase gene in HEK 293 cells.[16] This difference in transfection efficacy could be due to availability of an appropriate balance between protection of DNA and release for biological activity.

It could be deduced that high MW chitosans are better than low MW chitosans in improving the nanoparticles' stability, which is beneficial for the protection of DNA in the cellular endosomal/lysosomal compartments, but also inhibits dissociation of DNA once into the cytoplasm, or allows slow release, resulting in low or delayed expression.[23,26] On the other hand, nanoparticles prepared with chitosans of very low MW are not sufficiently stable and are incapable of protecting DNA, due to early dissociation, and therefore result in low or no transgene expression. Hence, a subtle balance must be achieved between extracellular DNA protection (better with high MW) versus efficient intracellular unpackaging (better with low MW) to obtain high transfection efficiency of chitosan.

9.2.2 Degree of deacetylation

DDA of chitosan represents the percentage of deacetylated primary amine groups along the molecular chain, which further determines the cationic charge density when chitosan is dissolved in acidic medium. Higher DDA yields increased positive charge, which allows a greater DNA binding capacity and cellular uptake. Furthermore, the DDA of chitosan also affects its degradation, solubility, and crystallinity.[14]

The size of chitosan nanoparticles has been observed to decrease from 181 to 239 nm with the decrease in DDA from

88 to 46%, while the zeta potential dropped from 22.2 mV to 9.6 mV.[9] Kiang et al. reported that DDA of chitosan significantly influences DNA binding, release, and gene transfection efficiency *in vitro* and *in vivo*.[27] For chitosans with a MW of 390 kDa, the charge ratio to achieve complete DNA complexation for DDA of 90%, 70%, and 62% was 3.3:1, 5.0:1, and 9.0:1, respectively.[27] Lowering of DDA decreases the overall *in vitro* gene expression levels in HEK293, HeLa, and SW756 cells, due to the instability of the particles in the presence of serum proteins. However, this instability, along with the increased chitosan degradation rate, contributes to higher luciferase transgene expression levels in muscles of female Balb/c mice, revealing the disparity that can result between *in vitro* and *in vivo* gene expression studies.[27] Also, Koping-Hoggard et al. suggested that the DDA of chitosan must be more than 65% in order to obtain stable complexes with pDNA that can efficiently transfect target cells *in vitro*.[26] While investigating the effect of MW and DDA on chitosan, Lavertu et al. suggested that maximum expression levels could be obtained by simultaneously lowering the MW and increasing DDA, or lowering DDA and increasing the MW, indicating a predominant role of particle stability, through co-operative electrostatic binding, in determining efficiency.[16]

In one of our studies, nanoparticles prepared at higher DDA were observed to bind at a greater level to cells for all tested time points.[17] A strong correlation between cell binding and cell uptake, with a notable increase in uptake for the high DDA (92% and 80%) chitosans as compared with lower DDA (72%) chitosans, was observed (Figure 9.3). Uptake data suggested that the number of cells positive for uptake peaked at around 8 hours, when nearly 100% of the cells contained internalized polyplexes.[17] Furthermore, the kinetics of nanoparticle decondensation in relation to

Published by Woodhead Publishing Limited, 2013

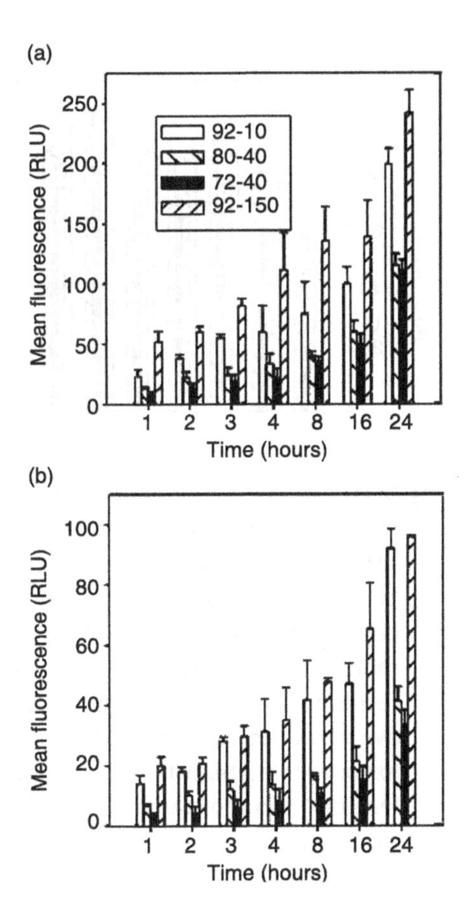

Figure 9.3 Kinetics of cellular binding and uptake of polyplexes prepared with chitosans of different DDA and MW.

HEK293 cells were incubated with fluorescent chitosan polyplexes for the indicated periods of time and analyzed by flow cytometry. Flow cytometry quantitative analysis of mean fluorescence per cell for polyplex (a) binding and (b) uptake and (c) of % cells with internalized polyplexes was performed following trypsinization and extensive washes, except for (a) cell binding, where cells were detached by enzyme-free cell dissociation buffer and analyzed directly.

(continued)

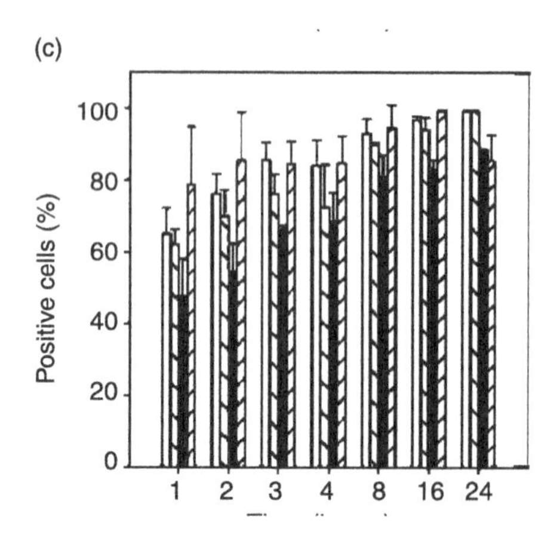

Figure 9.3 (*continued*) (b) Mean uptake levels per cell and (c) % positive cells were obtained from the same set of flow cytometry data. Graphs show that binding and uptake are time and DDA-dependent, with both 92% DDA chitosans binding more effectively than the lower DDA chitosans, resulting in increased uptake. Results are the average of three (n = 3) independent experiments ± SD, where each experiment included two replicates

Source: adapted from Thibault et al. 2010[17]

lysosomal sequestration and escape was found to be critically dependent on chitosan structure. The low DDA chitosan (72%) possesses weak stability that could lead to substantial and accelerated pDNA degradation in lysosomes, while high DDA chitosan (92% and 80%) provides protection against enzymatic digestion due to the ability to protect pDNA until lysosomal escape to the cytosol.[17]

9.2.3 N/P ratio

The N/P ratio in the case of chitosan nanoparticles is defined as the ratio of nitrogen (N) of chitosan to phosphate (P) of

Published by Woodhead Publishing Limited, 2013

DNA. The surface charge or zeta potential of the nanoparticles depends on the molar stoichiometry of chitosan to DNA, which later determines the capability of chitosan to efficiently condense pDNA and interact with negatively charged cell membranes, and ultimately its fate in terms of transfection efficiency.[9,25] The transfection efficiency of chitosan nanoparticles was observed to increase at charge ratios of 3 and 5, and decreased at higher charge ratios in SOJ cells.[11] This dependence was also verified with chitosan/pGL3 (N/P ratio 5) nanoparticles yielding high luciferase gene expression in human lung carcinoma, A549 cells.[24] The N/P ratios have been shown to influence the morphology of chitosan nanoparticles. Manipulation of N/P ratio in nanoparticles (average sizes 150–600 nm) resulted in different topological conformations including spherical, annular, toroidal, and globular morphologies.[18–20,28] The *in vitro* and *in vivo* transfection efficacies have also been correlated with the physical shape and morphology of nanoparticles.[18]

Cationic nanoparticles have been reported to bind to anionic microtubules or molecular motor proteins that facilitate movement towards the nuclear membrane along with the cytoskeletal network, resulting in enhanced cytoplasmic trafficking.[29] Moreover, an increase in the N/P ratio of the nanoparticles infers an increase in the chitosan concentration in the nano-complex. Increased amounts of chitosan in the nano-complexes may lead to a higher osmotic pressure in the endosomes that could favor increase in the efficiency of plasmid release.[26] In contrast, nano-complexes with a neutral surface charge tend to aggregate due to the absence of an inter-particle repulsive force.[30] However, there exists an optimal N/P ratio specific to the employed chitosan, due to the fact that use of too low a N/P ratio will yield physically unstable nano-complexes and poor transfection, while nano-complexes prepared at too high a N/P ratio with over-stability may also result in reduced transfection.[18,24,26]

The N/P ratio values for complete DNA condensation have been observed to be influenced by the chitosan structure. Evident interaction between MW and N/P ratio has been reported, where a high MW chitosan showed a higher transfection efficiency at a low N/P ratio, and a low MW chitosan needed a higher N/P ratio to completely form nano-complexes.[12,16,31] This could be due to optimal complexation and de-complexation of chitosan with DNA in the chitosan/DNA nanoparticles, prepared at an optimal MW of chitosan and N/P ratio. However, at the same MW, a lower DDA requires a higher N/P ratio to completely condense DNA, while at the same DDA a lower MW chitosan requires a higher N/P ratio.[16,26,27]

9.2.4 Chitosan salt form

Although chitosan and its derivatives have been extensively reported to be effective agents for transfection, they are only soluble in acidic solutions. However, chitosan salts such as chloride or lactate salts are often water-soluble and have been reported to possess improved transfection efficiency. Chitosan lactate and acetate of varying MW (20, 45, 200, and 460 kDa) were reported to have comparable transfection efficiencies and cell viabilities above 90%.[32] Further, chitosan lactate was observed to require a higher charge ratio in comparison with chitosan acetate to completely form complexes with DNA. This difference was attributed to the availability of different counter-ions (acetate ion and lactate ion) present in the medium that could interact with chitosan. Later, chitosan/DNA nanoparticles formulated with various chitosan salts, including chitosan hydrochloride (CHy), chitosan lactate (CLa), chitosan acetate (CAc), chitosan

Published by Woodhead Publishing Limited, 2013

aspartate (CAs), and chitosan glutamate (CGl), were investigated for transfection potential in CHO-K1 cells and were found to be dependent on the salt form.[12] The transfection efficiency had a tendency to increase as the N/P ratio was increased. CHy/DNA, CLa/DNA, CAc/DNA, CAs/DNA, and CGl/DNA complexes showed maximum transfection efficiencies at N/P ratios of 12, 12, 8, 6, and 6, respectively. All the salt forms have a transfection efficiency superior to that of chitosan base.[12]

9.2.5 Concentration of DNA

The transfection efficiency of chitosan/DNA nanoparticles also depends on the amount of DNA incorporated within the nanoparticles. As is evident from literature, the transfection efficiency increases with plasmid concentration up to an optimal point, after which the transfection remains constant or starts to decrease. Increase in plasmid concentration was observed to increase the diameter of the nanoparticles formed.[23] A greater increase in size was observed by increasing the plasmid concentration and formulating with a higher MW chitosan (102 kDa) than with a lower MW chitosan (32 kDa), inferring that it is possible to formulate complexes of a specific diameter by adjusting the plasmid concentration and chitosan MW.[23] Further, the transfection efficiency of nanoparticles prepared at a plasmid concentration of 200 μg/ml was significantly higher than those prepared at 50 and 100 μg/ml. In another study, increase in luciferase gene expression was observed with increasing the plasmid concentration from 0.5 to 2.5 μg/well.[31] However, the gene expression levels were saturated in epithelioma papulosum cyprini (EPC) cells as a result of a further increase in the DNA concentration to 5 μg/well. A later study suggested an increased level of transfection efficiency in primary

Published by Woodhead Publishing Limited, 2013

chondrocytes with an increase in the plasmid dosage.[33] Furthermore, the transfection efficiency decreased greatly when the plasmid dosage increased up to 16 and 32 μg/well. This decrease was attributed to the aggregation of nanoparticles with their increasing amount, resulting in impaired cell uptake.[33]

9.2.6 pH of transfection medium

The charge of chitosan/DNA nanoparticles, which dictates the transfection efficacy, depends on the concentration of DNA and chitosan as well as the pH and salt content of the suspension medium. The pKa of chitosan can be expressed as a linear function of its charge density with an intrinsic pKa (pK_0) of 6.7 and, as for any polyelectrolyte, this charge density dependence of pKa is reduced as ionic strength increases.[34] At 50% protonation, in the presence of 15 or 150 mM of NaCl, the pKa of the amino groups of chitosan is about 6.3 or 6.5, respectively, and the polymer's cationic charge density is greatly reduced by pH increases in the 5.5 to 7.5 region. For instance, increasing pH from 5.5 to 7.5 results in a decrease of chitosan amine protonation from about 75% to 10% in 15 mM NaCl or a decrease from about 90% to 10% in 150 mM NaCl, according to a recently developed molecular model of chitosan ionization.[34] Interestingly, the strongly polyanionic nature of DNA in chitosan/DNA nanoparticles facilitates the protonation of glucosamine units of chitosan due to proton transfer from the buffer to chitosan during complex formation, even at a high pH such as 7.4, at which chitosan would be largely uncharged in the absence of DNA.[35] We observed the zeta potential of chitosan/DNA nanoparticles in water to be 41.4 ± 5.1 mV, which generates sufficient electrostatic

Published by Woodhead Publishing Limited, 2013

repulsion to prevent aggregation of complexes during incubation times exceeding 120 min, was further reduced by suspension in PBS at pH 6.5, and even became negative at pH 7.4[15] (Table 9.1). This dependence of charge on pH was consistent with that previously reported, where electrostatically neutral particles were found in the pH range of 7.0–7.4 using an N/P ratio of 6, while the zeta potential became −20 mV at pH 8–8.5.

The transfection efficiency of chitosan/DNA nanoparticles in A549 cells was observed to be pH-dependent, with higher efficacy at pH 6.9 as compared with pH 7.6, due to the fact that nanoparticles at pH 6.9 are positively charged, and can bind with the negatively charged cells through electrostatic interaction.[24] Further, the transfection efficiency at pH 6.5 with chitosan (10 kDa) was reported to be higher than that at pH 7.1 in HEK 293 cells.[16] Furthermore, highest gene expression was obtained at pH 6.8 and 7.0; the transfection efficiency decreased dramatically when pH of the transfection medium increased to 7.4, which was explained by the

Table 9.1 Hydrodynamic diameter, polydispersity index and zeta potential (mean ± SD, $n = 3$) of chitosan/pDNA complexes in various media without serum

S.No.	Measurement medium	Hydrodynamic diameter (nm)	Polydispersity index	Zeta potential (mV)
1.	Double-distilled water	243±12	0.39±0.07	41.4±5.1
2.	10 mM Nacl	391±43.7	0.41±0.11	28±5.2
3.	150 mM Nacl	890±71.6	0.20±0.08	23±6.3
4.	PBS pH 6.5	911±39.6	0.14±0.07	11.4±4.1
5.	PBS pH 7.1	1213±84	0.10±0.06	4.5±1.0
6.	PBS pH 7.4	1244±135.2	0.19±0.5	−4.9±3.5

Published by Woodhead Publishing Limited, 2013

dissociation of free plasmid from the complex at higher pH.[11] On the contrary, in a more acidic pH, the transfection efficiency decreases due to very strong electrostatic interactions between the negatively charged DNA and the positively charged chitosan, which ultimately slows down DNA release.

In one of our studies, uptake of chitosan nanoparticles by HEK 293 cells was found to be pH-dependent, with maximum uptake occurring in medium at pH 6.5[15] (Figure 9.4). After 4 h of incubation at pH 6.5 with serum, almost 100% of cells internalized complexes and chitosan alone, whereas at pH 7.1 with serum only 55% of cells internalized complexes, while 100% internalized chitosan only. Further, the transfection efficiency, expressed as percentage of cells expressing EGFP, was observed to be 26.3% at pH 6.5, and then dropped considerably at higher pH values of 7.1 and 7.4 to 9.2% and 0.2%, respectively[15] (Figure 9.5). This was attributed to the fact that the pKa of the amino groups in chitosan is ~6.5; hence, in the transfection medium at pH 6.5, chitosan is expected to be highly protonated and the chitosan–DNA nanoparticles to be positively charged. High cationic charge of nanoparticles at pH 6.5 could result in a high non-specific affinity for negatively charged cell membranes and consequently high cell uptake, thereby producing higher transfection efficiency than at a higher pH of 7.1 or 7.4.[15]

9.2.7 Presence of serum

Development of gene delivery vectors that are stable even in the presence of serum is highly desirable for the practical applications of *in vivo* gene therapy. Numerous studies have proposed higher transfection efficiency for chitosan nanoparticles in the presence of serum rather than the absence of serum. The transfection efficiency of chitosan

Published by Woodhead Publishing Limited, 2013

nanoparticles in HeLa cells in the presence of 10% serum was reported to be higher than in the absence of serum, whereas PEI-mediated transfection decreased in the presence of 10% serum.[19] The influence of presence of 10–50% serum on the transfection efficiency of the chitosan/pGL3 nanoparticles was investigated in A549 cells, and it was shown that the presence of 10–20% serum resulted in improved gene transfer efficiency.[24] At a serum content of 20%, gene expression with chitosan/pGL3 nanoparticles was increased by two to threefold compared with that without serum, which was attributed to the increase in cell function by the addition of serum. However, further increase in serum content, up to 30–50%, resulted in a decrease of transfection efficiency, due to cell damage induced by addition of a high amount of serum.[24]

We observed serum-dependent uptake of chitosan nanoparticles by HEK 293, with maximum uptake occurring in medium at pH 6.5 supplemented with 10% FBS[15] (Figure 9.4). The uptake of chitosan nanoparticles as well as of chitosan alone was higher in the presence of serum at all the three pH values investigated, pH 6.5, 7.1, and 7.4. Further, the transfection efficiency was found to be 20 to 50% lower in the absence of serum than in the presence of serum for time points of 8 h, 12 h and 24 h at pH 6.5 or 7.4. The serum-deprived cells continue to cycle until they complete mitosis, whereupon they exit into the G0 state and further division stops.[36,37] The increase in transfection efficiency in the presence of serum could thus also be attributed to serum promoting cell division.

9.2.8 Effect of additives

Although low MW chitosans are more advantageous for clinical applications, with properties such as increased

Published by Woodhead Publishing Limited, 2013

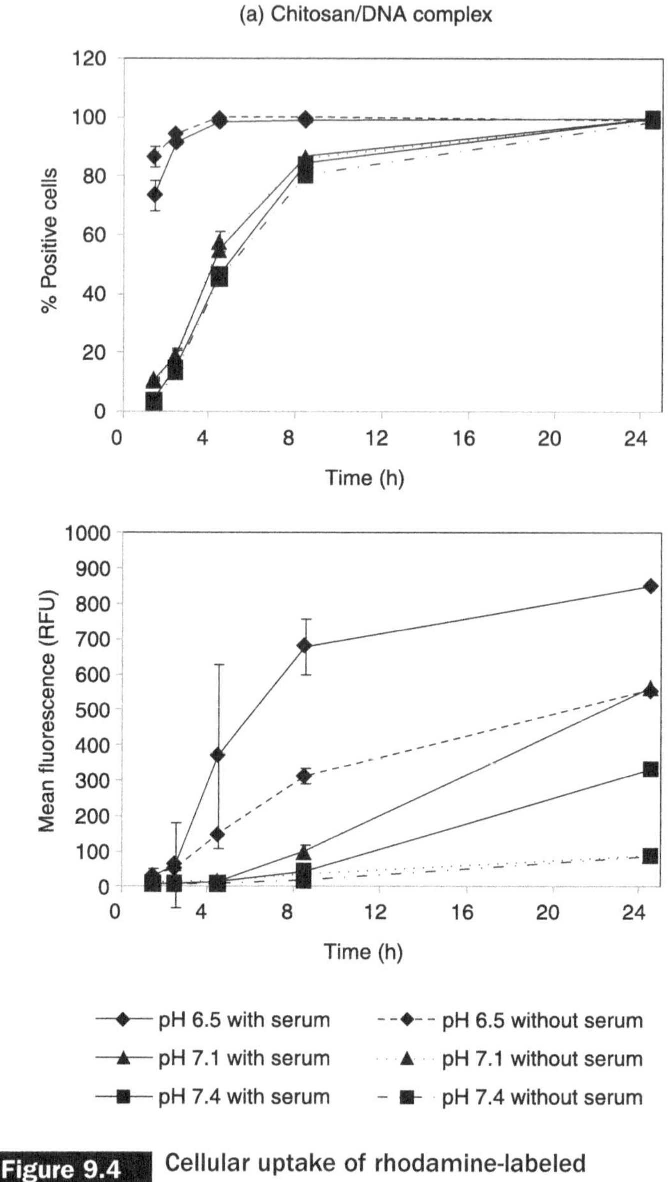

Figure 9.4 Cellular uptake of rhodamine-labeled chitosan/DNA complexes (a) and rhodamine-labeled chitosan

Published by Woodhead Publishing Limited, 2013

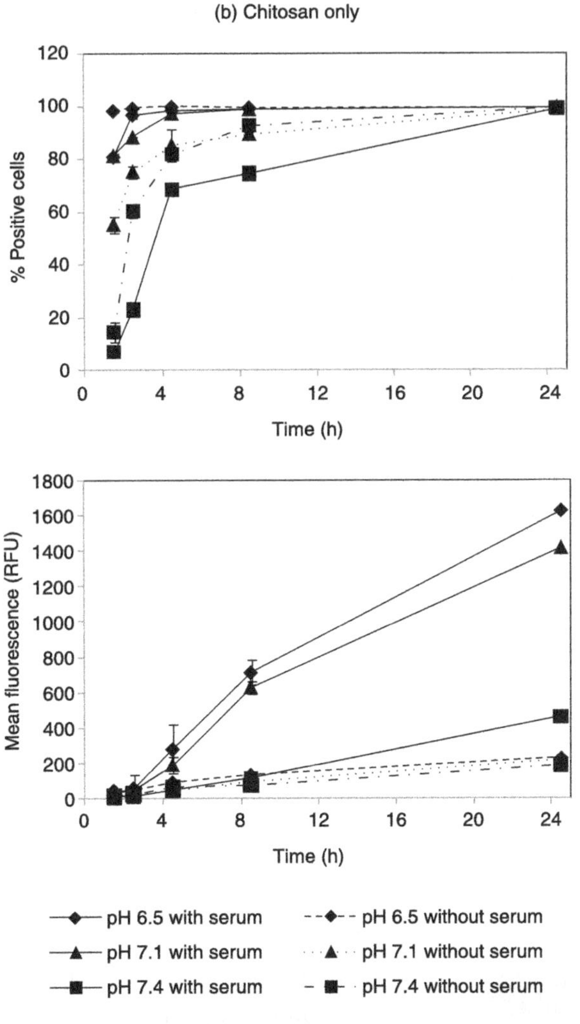

Figure 9.4 **(*continued*) (b) at different pH values**
HEK 293 cells were incubated with labeled complexes or chitosan at different pH values. After the stipulated time points, the percentage of cells with internalized label (top) and the mean fluorescence intensity of the label per cell (bottom) were determined by flow cytometry. Values are mean ± SD, $n = 3$

Source: adapted from Nimesh et al. 2010[15]

Published by Woodhead Publishing Limited, 2013

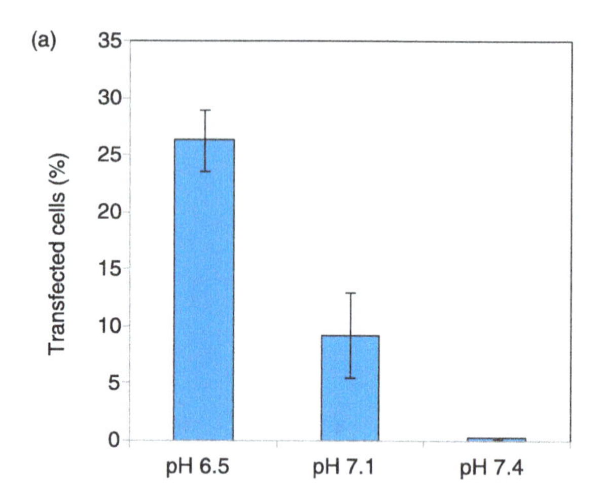

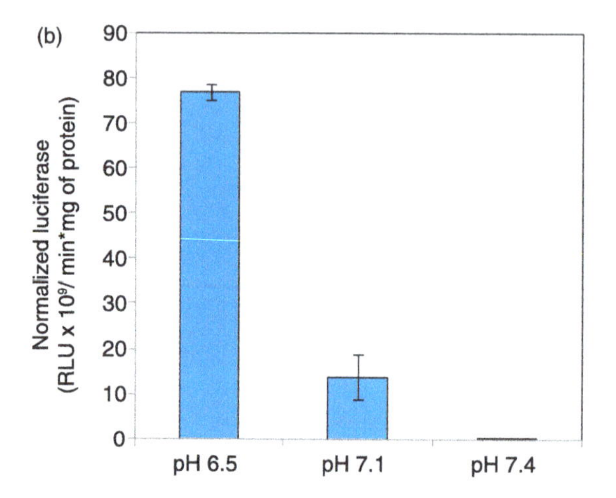

Figure 9.5 Transfection efficiency of chitosan/DNA complexes at different pH values

HEK 293 cells were transfected with chitosan/DNA complexes at different pH values. After 48 h (a) GFP expression was quantified using flow cytometry and expressed as percentage of cells transfected, while (b) the level of gene expression was determined by luminometry and expressed as RLU per min per mg of protein. Values are mean ± SD, $n = 3$

Source: adapted from Nimesh et al.[15]

Published by Woodhead Publishing Limited, 2013

solubility at physiological pH and improved DNA release, they suffer from low transfection efficacies as compared with high MW chitosans.[16] One approach for the improvement of low MW chitosan/DNA particle stability and transfection efficiency involves the association of low MW chitosan with an anionic biopolymer, such as alginate, prior to the addition of DNA. This allows the formation of a more stable polyplex and smaller nanoparticles than the ones formed with low MW chitosan (10 kDa) alone. The low MW chitosan/DNA/alginate (12–80 kDa) nanoparticles were able to transfect 293T cells with a higher efficiency than nanoparticles without alginate.[38] Another strategy consists of complexation of low MW chitosan with anionic HA before condensation of DNA. Nanoparticles prepared by electrostatic conjugation of chitosan with HA were found to be in the size range of 110–230 nm, with a positive zeta potential of +10 to +32 mV and a very high pDNA association efficiency of 87–99% (w/w).[39] Further, the nanoparticles exhibited low cytotoxicity and transfection levels up to 25% HEK 293 cells with GFP expression. Low MW chitosan (10–12 kDa)-HA nanoparticles also showed high levels of expression of secreted alkaline phosphatase in the human corneal epithelium cell model.[40] Additionally, topical administration of nanoparticles to rabbits demonstrated that these nanoparticles entered the corneal and conjunctival epithelial cells and then became assimilated by the cells. In another study, the low MW chitosan-HA nanoparticles had a size in the range of 100–235 nm and a zeta potential of −30 to +28 mV.[41] The results of the transfection studies showed that nanoparticles were able to provide high transfection levels (up to 15% of transfected cells) in human corneal epithelial (HCE) and normal human conjunctival (IOBA-NHC) cell lines, without affecting cell viability.[41] A similar study employed low MW chitosan (5 kDa) and HA (MW of

Published by Woodhead Publishing Limited, 2013

64 kDa) at a weight ratio of 4:1 to prepare nanoparticles with a size of ~146 nm with high transfection efficiency in 293T cells.[42] The MW of the chitosan used for preparing nanoparticles played a significant role in DNA release; the lower MW facilitated better DNA release and thereby increased transfection efficiency.[42]

9.2.9 Stability against polyanions

Nanoparticles with sufficient stability against degradative enzymes in the bloodstream should be consequently delivered to the target cells without loss of integrity of the entrapped therapeutic nucleic acid-based drugs. However, the presence of various proteoglycans, which are proteins covalently cross-linked with carboxylic or sulfated glycoaminoglycons (GAGs), appears as a possible barrier for the nanoparticles on the way to the target cells for cargo delivery.[43] The GAGs, including heparin sulfate, chondroitin sulfate, and HA, are highly anionic and may interact with the cationic nanoparticles, which may affect the integrity of the particles and the mobility of the nanoparticles in the tissue extracellular matrix, limiting their access to target cells.[44,45] Hence, nanoparticles prepared from chitosan should be stable against anions. Nanoparticles prepared from high MW chitosan (102 kDa) at a N/P ratio of 2 were observed to be extremely stable against exposure to SDS and heparin (sodium salt), which facilitated *in vivo* application.[23]

9.2.10 Addition of chitosan

Though chitosan has been employed as the main vector in several studies involving DNA delivery, it can also be used as a coating material to offer a flexible technology platform.

Published by Woodhead Publishing Limited, 2013

The use of PLGA for DNA delivery applications is limited by its negative surface charge and acidic degradation products. The incorporation of chitosan into the PLGA matrix could inhibit degradation of the matrix, thereby blocking the release of DNA from the polymer. Chitosan-coated nanoparticles are characterized by a strongly positive zeta potential, in contrast to the non-coated PLGA nanoparticles. Therefore, aggregation of these nanoparticles is prevented as they are stabilized by strong electrostatic repulsion. Moreover, this positive charge may contribute towards improved cell attachment in fibroblast cells.[46] The chitosan-coated PLGA nanoparticles were established as a flexible and efficient delivery system for AS-ODNs to lung cancer cells.[25] Chitosan was shown to enhance the efficacy of emulsion as a gene delivery vehicle. Chitosan mediated pre-condensation of pDNA and enhanced transfection efficiency of the emulsion *in vitro* in HepG2 cells. The *in vivo* data suggested that the chitosan-enhanced emulsion complexes lasted longer in the organs of mice compared with those complexes without chitosan.[2] Further, coating with chitosan was observed to improve the loading efficiency of the nucleic acids and significantly reduced the initial burst of nucleic acid, resulting in sustained release.[47] Furthermore, by coating liposomes with chitosan, the luciferase gene expression levels were slightly reduced, possibly due to the increased instability of the chitosan-loaded liposomes.[48]

9.2.11 Preparation techniques of chitosan nanoparticles

Several strategies have been proposed for preparation of chitosan nanoparticles, such as coacervation, ionic gelation, covalent cross-linking, and desolvation.[49–51] The variation in preparation techniques can significantly alter the size of the

Published by Woodhead Publishing Limited, 2013

chitosan/DNA nanoparticles produced, and hence the overall transfection efficiency.[52–54] However, limited information is available in this area and further investigations are needed. Nanoparticles should be synthesized keeping in mind the route of administration.

9.2.12 Effect of route of administration

Route of administration to attain acceptable bioavailability represents a significant challenge for gene delivery. Usually DNA is delivered by parenteral administration, due to its physico-chemical characteristics, while oral administration has been less explored. However, chitosan has mucoadhesive properties, which facilitate its application in mucosal gene delivery through oral, intranasal, and intratracheal routes.[55,56] Chitosan has been widely employed for delivery of oral vaccines, as it has been demonstrated to have the additional advantage of promoting adhesion and absorption across a mucous surface. Chitosan has been reported to successfully deliver a reporter gene orally to enterocytes, Peyer's patches, and mesenteric lymph nodes in female New Zealand white rabbits.[23] In another study, oral gene immunization for reducing the food-induced anaphylactic response was observed in mice using a chitosan-based gene delivery system.[57] Also, orally administered chitosan/DNA nanoparticles were reported to stimulate an immune response to the principal peanut allergen Arah-2 in Swiss albino mice.[58] Later, high transfection efficiencies of pDNA were retained in Japanese flounder after oral administration by encapsulating pDNA in chitosan microspheres through an emulsion-based methodology, suggesting that chitosan microspheres may be promising carriers for oral pDNA vaccines.[59]

Intranasal administration is one of the most promising potential routes for gene therapy. Mice vaccinated with

Published by Woodhead Publishing Limited, 2013

chitosan nanoparticles encapsulating a cocktail of RSV cDNAs exhibited a significant reduction in RSV titer and antigen load, as compared with untreated and naked DNA-administered controls, after intranasal administration.[60] Another study reported that intranasal immunization with chitosan nanoparticles induced peptide- and virus-specific CTL responses in BALB/c mice that were comparable to those induced via intradermal immunization.[61] Additionally, pulmonary administration of the DNA plasmid incorporated in chitosan nanoparticles was shown to induce increased levels of IFN-γ secretion compared with plasmid DNA in solution or the intramuscular immunization route in HLA-A2 transgenic mice.[62] Moreover, chitosan/DNA nanoparticle powders were demonstrated to possess higher transfection potency than solutions containing the same amount of DNA via pulmonary administration in mice.[63]

9.2.13 Cell type

An efficient gene delivery system is required to condense DNA, transport the gene into the cell, and facilitate its release, thereby leading to gene expression and subsequent protein synthesis.[64] Since chitosan-mediated transfection has been observed to depend on the cell type, it is a prerequisite to test a gene carrier on different cell lines, including cells that resemble those that will be targeted. The composition of cellular membranes varies among cellular types and may promote or inhibit the binding of the nanoparticles and subsequent internalization.[64] Chitosan/DNA nanoparticles have been reported to transfect various cell types, including HeLa, HEK 293, A549, and COS-1 cells with an unknown mechanism.[27] HEK 293 cells were reported to be more efficiently transfected than HeLa, M69 mesenchymal, and MG63 cells with chitosan nanoparticles.[64,65] Another study

Published by Woodhead Publishing Limited, 2013

showed that HEK293 cells were transfected with chitosan nanoparticles, whereas HeLa and Swiss 3T3 cells were resistant to transfection.[66] Also, gene expression in HEK 293 cells with ultrapure and non-toxic chitosan nanoparticles was comparable to that obtained with PEI, while in more differentiated epithelial cells such as HT-1080 and Caco-2 it was less efficient.[26]

9.3 Conclusions

Chitosan represents one of the most widely investigated cationic polymers for gene delivery. The transfection efficiency is driven by a series of formulation parameters, such as the MW, DDA of chitosan, N/P ratio of chitosan/DNA nanoparticles, chitosan salt form, pH, serum, etc. Keeping in mind all the formulation-based investigations, it appears that there exists a range of intermediate values of MW and DDA which form complexes of optimal stability and transfect efficiently. As is evident from literature, the future prospects of chitosan-mediated gene delivery could be promising, despite the current problems.

9.4 References

1. Shu, X.Z. and Zhu, K.J. (2002) The influence of multivalent phosphate structure on the properties of ionically cross-linked chitosan films for controlled drug release. *Eur J Pharm Biopharm*, **54**, 235–43.
2. Lee, M.K., Chun, S.K., Choi, W.J., Kim, J.K., Choi, S.H., et al. (2005) The use of chitosan as a condensing agent to enhance emulsion-mediated gene transfer. *Biomaterials*, **26**, 2147–56.

Published by Woodhead Publishing Limited, 2013

3. Hejazi, R. and Amiji, M. (2003) Chitosan-based gastrointestinal delivery systems. *J Control Release*, **89**, 151–65.
4. Yi, H., Wu, L.Q., Bentley, W.E., Ghodssi, R., Rubloff, G.W., et al. (2005) Biofabrication with chitosan. *Biomacromolecules*, **6**, 2881–94.
5. Aiba, S. (1992) Studies on chitosan: 4. Lysozymic hydrolysis of partially N-acetylated chitosans. *Int J Biol Macromol*, **14**, 225–8.
6. Zhang, H. and Neau, S.H. (2002) In vitro degradation of chitosan by bacterial enzymes from rat cecal and colonic contents. *Biomaterials*, **23**, 2761–6.
7. Escott, G.M. and Adams, D.J. (1995) Chitinase activity in human serum and leukocytes. *Infect Immun*, **63**, 4770–3.
8. van der Lubben, I.M., Verhoef, J.C., Borchard, G. and Junginger, H.E. (2001) Chitosan and its derivatives in mucosal drug and vaccine delivery. *Eur J Pharm Sci*, **14**, 201–7.
9. Huang, M., Fong, C.W., Khor, E. and Lim, L.Y. (2005) Transfection efficiency of chitosan vectors: effect of polymer molecular weight and degree of deacetylation. *J Control Release*, **106**, 391–406.
10. Mumper, R.J., Wang, J., Claspell, J.M. and Rolland, A.P. (1995) Novel polymeric condensing carriers for gene delivery. *Proceedings of the International Symposium on Controlled Release Bioactive Materials*, **22**, 178–9.
11. Ishii, T., Okahata, Y. and Sato, T. (2001) Mechanism of cell transfection with plasmid/chitosan complexes. *Biochim Biophys Acta*, **1514**, 51–64.
12. Weecharangsan, W., Opanasopit, P., Ngawhirunpat, T., Apirakaramwong, A., Rojanarata, T., et al. (2008) Evaluation of chitosan salts as non-viral gene vectors in CHO-K1 cells. *Int J Pharm*, **348**, 161–8.

Published by Woodhead Publishing Limited, 2013

13. Koping-Hoggard, M., Varum, K.M., Issa, M., Danielsen, S., Christensen, B.E., et al. (2004) Improved chitosan-mediated gene delivery based on easily dissociated chitosan polyplexes of highly defined chitosan oligomers. *Gene Ther*, **11**, 1441–52.
14. Aiba, S. (1989) Studies on chitosan: 2. Solution stability and reactivity of partially N-acetylated chitosan derivatives in aqueous media. *Int J Biol Macromol*, **11**, 249–52.
15. Nimesh, S., Thibault, M.M., Lavertu, M. and Buschmann, M.D. (2010) Enhanced gene delivery mediated by low molecular weight chitosan/DNA complexes: effect of pH and serum. *Mol Biotechnol*, **46**, 182–96.
16. Lavertu, M., Methot, S., Tran-Khanh, N. and Buschmann, M.D. (2006) High efficiency gene transfer using chitosan/DNA nanoparticles with specific combinations of molecular weight and degree of deacetylation. *Biomaterials*, **27**, 4815–24.
17. Thibault, M., Nimesh, S., Lavertu, M. and Buschmann, M.D. (2010) Intracellular trafficking and decondensation kinetics of chitosan-pDNA polyplexes. *Mol Ther*, **18**, 1787–95.
18. Koping-Hoggard, M., Mel'nikova, Y.S., Varum, K.M., Lindman, B. and Artursson, P. (2003) Relationship between the physical shape and the efficiency of oligomeric chitosan as a gene delivery system in vitro and in vivo. *J Gene Med*, **5**, 130–41.
19. Erbacher, P., Zou, S., Bettinger, T., Steffan, A.M. and Remy, J.S. (1998) Chitosan-based vector/DNA complexes for gene delivery: biophysical characteristics and transfection ability. *Pharm Res*, **15**, 1332–9.
20. Danielsen, S., Strand, S., de Lange Davies, C. and Stokke, B.T. (2005) Glycosaminoglycan destabilization

Published by Woodhead Publishing Limited, 2013

of DNA-chitosan polyplexes for gene delivery depends on chitosan chain length and GAG properties. *Biochim Biophys Acta*, **1721**, 44–54.

21. Li, H.Y. and Birchall, J. (2006) Chitosan-modified dry powder formulations for pulmonary gene delivery. *Pharm Res*, **23**, 941–50.
22. Kawamata, Y., Nagayama, Y., Nakao, K., Mizuguchi, H., Hayakawa, T., et al. (2002) Receptor-independent augmentation of adenovirus-mediated gene transfer with chitosan in vitro. *Biomaterials*, **23**, 4573–9.
23. MacLaughlin, F.C., Mumper, R.J., Wang, J., Tagliaferri, J.M., Gill, I., et al. (1998) Chitosan and depolymerized chitosan oligomers as condensing carriers for in vivo plasmid delivery. *J Control Release*, **56**, 259–72.
24. Sato, T., Ishii, T. and Okahata, Y. (2001) In vitro gene delivery mediated by chitosan. effect of pH, serum, and molecular mass of chitosan on the transfection efficiency. *Biomaterials*, **22**, 2075–80.
25. Nafee, N., Taetz, S., Schneider, M., Schaefer, U.F. and Lehr, C.M. (2007) Chitosan-coated PLGA nanoparticles for DNA/RNA delivery: effect of the formulation parameters on complexation and transfection of antisense oligonucleotides. *Nanomedicine*, **3**, 173–83.
26. Koping-Hoggard, M., Tubulekas, I., Guan, H., Edwards, K., Nilsson, M., et al. (2001) Chitosan as a nonviral gene delivery system. Structure-property relationships and characteristics compared with polyethylenimine in vitro and after lung administration in vivo. *Gene Ther*, **8**, 1108–21.
27. Kiang, T., Wen, J., Lim, H.W. and Leong, K.W. (2004) The effect of the degree of chitosan deacetylation on the efficiency of gene transfection. *Biomaterials*, **25**, 5293–301.

Published by Woodhead Publishing Limited, 2013

28. Liu, W., Sun, S., Cao, Z., Zhang, X., Yao, K., et al. (2005) An investigation on the physicochemical properties of chitosan/DNA polyelectrolyte complexes. *Biomaterials*, **26**, 2705–11.
29. Jeong, J.H., Kim, S.W. and Park, T.G. (2007) Molecular design of functional polymers for gene therapy. *Prog Polym Sci*, **32**, 1239–74.
30. De Smedt, S.C., Demeester, J. and Hennink, W.E. (2000) Cationic polymer based gene delivery systems. *Pharm Res*, **17**, 113–26.
31. Romoren, K., Pedersen, S., Smistad, G., Evensen, O. and Thu, B.J. (2003) The influence of formulation variables on in vitro transfection efficiency and physicochemical properties of chitosan-based polyplexes. *Int J Pharm*, **261**, 115–27.
32. Weecharangsan, W., Opanasopit, P., Ngawhirunpat, T., Rojanarata, T. and Apirakaramwong, A. (2006) Chitosan lactate as a nonviral gene delivery vector in COS-1 cells. *AAPS PharmSciTech*, **7**, 66.
33. Zhao, X., Yu, S.B., Wu, F.L., Mao, Z.B. and Yu, C.L. (2006) Transfection of primary chondrocytes using chitosan-pEGFP nanoparticles. *J Control Release*, **112**, 223–8.
34. Filion, D., Lavertu, M. and Buschmann, M.D. (2007) Ionization and solubility of chitosan solutions related to thermosensitive chitosan/glycerol-phosphate systems. *Biomacromolecules*, **8**, 3224–34.
35. Ma, P.L., Lavertu, M., Winnik, F.M. and Buschmann, M.D. (2009) New insights into chitosan-DNA interactions using isothermal titration microcalorimetry. *Biomacromolecules*, **10**, 1490–9.
36. Pardee, A.B. (1974) A restriction point for control of normal animal cell proliferation. *Proc Natl Acad Sci USA*, **71**, 1286–90.

37. Pardee, A.B. (1989) G1 events and regulation of cell proliferation. *Science*, **246**, 603–8.
38. Douglas, K.L., Piccirillo, C.A. and Tabrizian, M. (2006) Effects of alginate inclusion on the vector properties of chitosan-based nanoparticles. *J Control Release*, **115**, 354–61.
39. de la Fuente, M., Seijo, B. and Alonso, M.J. (2008) Design of novel polysaccharidic nanostructures for gene delivery. *Nanotechnology*, **19**, 075105.
40. de la Fuente, M., Seijo, B. and Alonso, M.J. (2008) Bioadhesive hyaluronan-chitosan nanoparticles can transport genes across the ocular mucosa and transfect ocular tissue. *Gene Ther*, **15**, 668–76.
41. de la Fuente, M., Seijo, B. and Alonso, M.J. (2008) Novel hyaluronic acid-chitosan nanoparticles for ocular gene therapy. *Invest Ophthalmol Vis Sci*, **49**, 2016–24.
42. Duceppe, N. and Tabrizian, M. (2009) Factors influencing the transfection efficiency of ultra low molecular weight chitosan/hyaluronic acid nanoparticles. *Biomaterials*, **30**, 2625–31.
43. Ruponen, M., Honkakoski, P., Ronkko, S., Pelkonen, J., Tammi, M., et al. (2003) Extracellular and intracellular barriers in non-viral gene delivery. *J Control Release*, **93**, 213–17.
44. Ruponen, M., Yla-Herttuala, S. and Urtti, A. (1999) Interactions of polymeric and liposomal gene delivery systems with extracellular glycosaminoglycans: physicochemical and transfection studies. *Biochim Biophys Acta*, **1415**, 331–41.
45. Pitkanen, L., Ruponen, M., Nieminen, J. and Urtti, A. (2003) Vitreous is a barrier in nonviral gene transfer by cationic lipids and polymers. *Pharm Res*, **20**, 576–83.
46. Nie, H., Lee, L.Y., Tong, H. and Wang, C.H. (2008) PLGA/chitosan composites from a combination of spray

Published by Woodhead Publishing Limited, 2013

drying and supercritical fluid foaming techniques: new carriers for DNA delivery. *J Control Release*, **129**, 207–14.

47. Tahara, K., Sakai, T., Yamamoto, H., Takeuchi, H. and Kawashima, Y. (2008) Establishing chitosan coated PLGA nanosphere platform loaded with wide variety of nucleic acid by complexation with cationic compound for gene delivery. *Int J Pharm*, **354**, 210–16.
48. Colonna, C., Conti, B., Genta, I. and Alpar, O.H. (2008) Non-viral dried powders for respiratory gene delivery prepared by cationic and chitosan loaded liposomes. *Int J Pharm*, **364**, 108–18.
49. Mao, H.Q., Roy, K., Troung-Le, V.L., Janes, K.A., Lin, K.Y., et al. (2001) Chitosan-DNA nanoparticles as gene carriers: synthesis, characterization and transfection efficiency. *J Control Release*, **70**, 399–421.
50. Berthold, A., Cremer, K. and Kreuter, J. (1996) Preparation and characterization of chitosan microspheres as drug carrier for prednisolone sodium phosphate as model for anti-inflammatory drugs. *J Control Release*, **39**, 17–25.
51. Hamidi, M., Azadi, A. and Rafiei, P. (2008) Hydrogel nanoparticles in drug delivery. *Adv Drug Deliv Rev*, **60**, 1638–49.
52. Guang Liu, W. and De Yao, K. (2002) Chitosan and its derivatives—a promising non-viral vector for gene transfection. *J Control Release*, **83**, 1–11.
53. Borchard, G. (2001) Chitosans for gene delivery. *Adv Drug Deliv Rev*, **52**, 145–50.
54. Mansouri, S., Lavigne, P., Corsi, K., Benderdour, M., Beaumont, E., et al. (2004) Chitosan-DNA nanoparticles as non-viral vectors in gene therapy: strategies to improve transfection efficacy. *Eur J Pharm Biopharm*, **57**, 1–8.

Published by Woodhead Publishing Limited, 2013

55. Kim, T.-H., Jiang, H.-L., Jere, D., Park, I.-K., Cho, M.-H., et al. (2007) Chemical modification of chitosan as a gene carrier in vitro and in vivo. *Prog Polym Sci*, **32**, 726–53.
56. Dang, J.M. and Leong, K.W. (2006) Natural polymers for gene delivery and tissue engineering. *Adv Drug Deliv Rev*, **58**, 487–99.
57. Roy, K., Mao, H.Q., Huang, S.K. and Leong, K.W. (1999) Oral gene delivery with chitosan–DNA nanoparticles generates immunologic protection in a murine model of peanut allergy. *Nat Med*, **5**, 387–91.
58. Guliyeva, U., Oner, F., Ozsoy, S. and Haziroglu, R. (2006) Chitosan microparticles containing plasmid DNA as potential oral gene delivery system. *Eur J Pharm Biopharm*, **62**, 17–25.
59. Tian, J., Yu, J. and Sun, X. (2008) Chitosan microspheres as candidate plasmid vaccine carrier for oral immunisation of Japanese flounder (Paralichthys olivaceus). *Vet Immunol Immunopathol*, **126**, 220–9.
60. Kumar, M., Behera, A.K., Lockey, R.F., Zhang, J., Bhullar, G., et al. (2002) Intranasal gene transfer by chitosan-DNA nanospheres protects BALB/c mice against acute respiratory syncytial virus infection. *Hum Gene Ther*, **13**, 1415–25.
61. Iqbal, M., Lin, W., Jabbal-Gill, I., Davis, S.S., Steward, M.W., et al. (2003) Nasal delivery of chitosan-DNA plasmid expressing epitopes of respiratory syncytial virus (RSV) induces protective CTL responses in BALB/c mice. *Vaccine*, **21**, 1478–85.
62. Bivas-Benita, M., van Meijgaarden, K.E., Franken, K.L., Junginger, H.E., Borchard, G., et al. (2004) Pulmonary delivery of chitosan-DNA nanoparticles enhances the immunogenicity of a DNA vaccine encoding HLA-A*0201-restricted T-cell epitopes of Mycobacterium tuberculosis. *Vaccine*, **22**, 1609–15.

Published by Woodhead Publishing Limited, 2013

63. Okamoto, H., Sakakura, Y., Shiraki, K., Oka, K., Nishida, S., et al. (2005) Stability of chitosan–pDNA complex powder prepared by supercritical carbon dioxide process. *Int J Pharm*, **290**, 73–81.
64. Corsi, K., Chellat, F., Yahia, L. and Fernandes, J.C. (2003) Mesenchymal stem cells, MG63 and HEK293 transfection using chitosan-DNA nanoparticles. *Biomaterials*, **24**, 1255–64.
65. Leong, K.W., Mao, H.Q., Truong-Le, V.L., Roy, K., Walsh, S.M., et al. (1998) DNA-polycation nanospheres as non-viral gene delivery vehicles. *J Control Release*, **53**, 183–93.
66. Dastan, T. and Turan, K. (2004) In vitro characterization and delivery of chitosan-DNA microparticles into mammalian cells. *J Pharm Pharm Sci*, **7**, 205–14.

Published by Woodhead Publishing Limited, 2013

10

Polyethylenimine nanoparticles

DOI: 10.1533/9781908818645.197

Abstract: Polyethylenimine is considered as the gold standard of gene transfection; is one of the largely investigated cationic polymers for gene delivery. PEI possesses high cation density (a positive charge per 43 Da) and depending on the linkage of the repeating ethylenimine units, it occurs as branched or linear morphological isomers. Branched PEI is synthesized by acid-catalyzed polymerization of aziridine whereas linear PEI is prepared via ring opening polymerization of 2-ethyl-2-oxazoline followed by hydrolysis. LPEI contains secondary amines in its backbone except the terminal primary groups. On the other hand, BPEI contains primary, secondary and tertiary amino groups at the estimated ratio of 1:2:1. PEIs, ranging from low MW to higher ones, have extensively been exploited as effective gene delivery vehicles. It has been earlier reported that of linear and branched PEIs, the latter ones are more toxic and less efficient for transfection, particularly at higher N/P ratios. The present chapter provides an exhaustive detail of the studies exploring PEI for *in vitro* and *in vivo* gene delivery.

Keywords: polyethylenimine, branched PEI, linear PEI, polyethylene glycol, PEGylated PEI, alkylation, acetylation, folate.

Published by Woodhead Publishing Limited, 2013

10.1 Introduction

A wide variety of linear, branched cationic polymers and copolymers have been tested in terms of their efficacy and suitability for *in vitro* transfection. Unfortunately, no system has emerged as a versatile vector for gene delivery.[1] Studies regarding the mechanism of condensation of polymer structure with DNA and their biological performance, such as toxicity and transfection efficiency, are limited. Because of this, the development of newer systems relies on empirical approaches rather than on a rational design. However, the results from transfection experiments with PEI have been encouraging from the very beginning. PEI, often considered as the gold standard of gene transfection, is one of the most widely explored cationic polymers for gene delivery. Since the first successful application by Behr et al. of PEI-mediated ODN delivery, it has been further derivatized to improve the physico-chemical and biological properties of polyplexes.[2,3] High cation density of PEI (one positive charge per 43 Da, which is the monomer's MW) also contributes to the formation of highly condensed particles by interacting with nucleic acids. Depending on the linkage of the repeating ethylenimine units, PEI occurs as branched or linear morphological isomers. Branched PEI is synthesized by acid-catalyzed polymerization of aziridine, whereas linear PEI is prepared via ring opening polymerization of 2-ethyl-2-oxazoline followed by hydrolysis (Figure 10.1).[4,5] LPEI contains secondary amines in its backbone except for the terminal primary groups. On the other hand, BPEI contains primary, secondary and tertiary amino groups at the estimated ratio of 1:2:1.[6] The different types of amine group have different pKa values and could be protonated at different levels at a given pH. This confers on PEI a superior buffering capacity over a wide pH range. PEI uses the "proton sponge" mechanism to promote the release of endocytosed polyplexes from the endosomes.[2,7–9]

Published by Woodhead Publishing Limited, 2013

PEIs, ranging from low to higher MW, have been extensively exploited as effective gene delivery vehicles.[2,10–13] The transfection efficiency of PEI is directly related to the size of the polymer and the charge – associated with cytotoxicity.[2,14–16] The MWs of PEI being investigated ranged from 1 to 1600 kDa.[17,18] Moreover, transfection data from L929 cells suggested that the most suitable MW of PEI for gene delivery ranges between 5 and 25 kDa.[14] Although the higher MW PEI is associated with higher transfection efficiency, it has also been shown to result in increased cytotoxicity due to cell-surface aggregation of the polymer.[14] On the other hand, low MW PEI is less toxic but is also less efficient in gene delivery. Also, low MW PEIs possess a lower amount of positive charge within one molecule, which makes it difficult to condense the

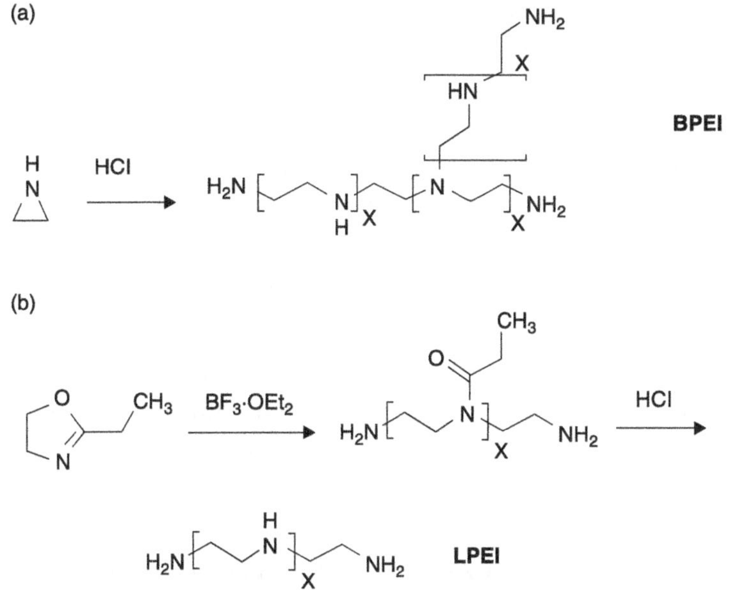

Figure 10.1 Synthesis of PEI by (a) acid-polymerization of aziridine to yield BPEI and (b) ring-opening polymerization of 2-ethyl-2-oxazoline followed by hydrolysis to yield LPEI

Published by Woodhead Publishing Limited, 2013

negatively charged DNA. On the other hand, if the surface charge of the complexes is too low, it is almost impossible for them to induce cellular uptake through charge-mediated interactions.[15] However, the high cationic charge density of PEI condenses the negatively charged DNA efficiently into small complexes and protects it from nuclease degradation.[14] Low MW PEI (5 kDa) grafted with PEG (5 kDa) exhibited higher transfection efficiency in both bronchial and alveolar cells than high MW PEI (25 kDa) grafted PEG polyplexes.[19]

It has been reported that branched PEIs are more toxic and less efficient for transfection than linear PEIs, particularly at higher N/P ratios.[20] BPEI have been shown to complex strongly with DNA and form smaller complexes than LPEI.[20,21] The condensation behavior of LPEI is significantly influenced by the preparation buffer conditions, whereas that of BPEI remains unaffected.[20] For instance, LPEI (22 kDa) condenses DNA in low ionic strength 5% glucose solution to 30–60 nm complexes, which increase to 1 μm in a high ionic strength solution.[2,21] Also, the *in vitro* transfection efficiency of LPEI/DNA complexes was appreciably higher than that of BPEI (800 Da)/DNA and BPEI (25 kDa)/DNA complexes when complexes were prepared in a salt-containing buffer.[20] Further, transfection studies done *in vivo* showed that LPEI/DNA complexes prepared in high salt conditions were 100-fold less active than those formed in low salt conditions. These data strongly suggested that efficient transgene expression depends on the size of the PEI/DNA complexes.

10.2 Derivatives of PEI for *in vitro* and *in vivo* gene delivery

An exhaustive variety of chemical modifications have been introduced to the PEI structure in order to improve

Published by Woodhead Publishing Limited, 2013

transfection efficiency of the polymer. The modifications have been targeted either to mask the surface charge to reduce toxicity or for attachment of ligands for targeted gene delivery.

10.2.1 Conjugation of PEG to PEI

In order to improve the *in vivo* half-life and cell viability of PEI, the positive charge is masked by coating the surface with neutral hydrophilic polymers such as PEG. Steric stabilization is achieved by creating a "brush" layer of hydrophilic polymer on the surface of PEI/DNA polyplexes, thereby decreasing self and non-self non-specific interactions. One of the first studies investigated the *in vitro* and *in vivo* properties of (800 kDa) complexes before and after covalent coupling of PEG (2 kDa). The majority of Tf-PEI/DNA and Tf-PEI/PEG/DNA complexes generated under low salt conditions (HEPES-buffered glucose) appeared as spherical particles with a diameter of 40 nm. Non-PEGylated complexes formed aggregates with plasma proteins; in contrast, PEG-modified complexes remained small and un-aggregated. *In vitro* transfection data in human cell lines K562 and Neuro2a suggested that PEGylation did not interfere with the transfection activity. Further, the PEGylation of complexes was observed to improve the retention time in blood after tail vein injection as compared with unmodified complexes.

Various synthetic strategies have been proposed for conjugation of PEG to PEI. Coupling is usually a two-step procedure, whereby PEG can be activated with either epoxide or isocyanate groups, followed by reaction with the amino groups of PEI.[22,23] Also, commercially available NHS-activated PEG can be used to couple with PEI.[24] Furthermore, bifunctional NHS-activated PEG with a vinyl sulfone group on the opposite end allows further functionalization of the

Published by Woodhead Publishing Limited, 2013

PEI-PEG block copolymer with targeting moieties such as arginine-glycine-aspartic acid (RGD) peptides to target integrin receptors on endothelial cells or galactose to target hepatocytes.[25,26] PEG chains of varying lengths have been employed to modify BPEI (MW 2–25 kDa) and LPEI (MW 22 kDa).[23,27,28] Investigations revealed that the degree of PEGylation and the MW of PEG strongly influence the properties of the resulting PEG-PEI conjugates.[23] It is noteworthy that all studies involving PEG-PEI/DNA complexes were performed with PEIs that were PEGylated before generation of the DNA complexes. This could probably be attributed to less efficient DNA condensation with PEGylated PEI in comparison to non-modified PEIs.[23,27] The covalent conjugation of PEI with PEG reduced the positive surface charge (zeta potential) of the polyplexes, whereas it only marginally affected their size.[23,27,28] The shielding of the polyplexes allowed increased half-life in the blood circulation. Further, the PEGylation of PEI/DNA polyplexes also reduced the non-specific ionic interactions between polyplexes and target cells.[27] However, some studies reported efficient gene transfer despite PEGylation.[23] Furthermore, the PEG-PEI conjugates are observed to be less cytotoxic than the non-modified polymers. This could be due to the fact that fewer PEG-PEI/DNA complexes are taken up by the cells as a result of reduced non-specific ionic interactions. Additionally, the PEG chains significantly enhanced the solubility of the PEI/DNA complexes. Polyplexes of LPEI/DNA at PEI concentration of 350 μg/ml and N/P ratio of 12 resulted in formation of precipitates, while PEG-PEI/DNA complexes with up to 1500 μg/ml of DNA could be easily generated.[27]

To reveal the details of dependence of PEGylation on PEI transfection efficacy, polymers were synthesized by grafting PEG (2 kDa) to PEI (25 kDa) at multiple ratios.[29] Increased PEG substitution was shown to lower toxicity and pDNA-

Published by Woodhead Publishing Limited, 2013

binding on a per total polymer weight basis, but not on a per PEI-backbone weight basis. Further, DLS studies indicated inhibition of aggregation at high PEGylation of polyplexes in the presence of serum. Plasmid uptake and transgene expression were found to have a complex relationship with PEG substitution, depending on the polymer/pDNA weight ratio. Furthermore, PEGylation generally decreased the transfection efficacy of PEI, but, under ideal conditions of PEG substitution and polymer/pDNA ratio, PEGylation provided more effective carrier formulations than the native PEI itself.

Although *in vivo* studies of PEG-PEI copolymers have shown increased blood circulation time and reduced toxicity, no gene expression was detected with doses of 25 μg pDNA in mice.[30] This was attributed to reduced interaction with the cell membrane due to decreased surface charge, hindering the first step of the intracellular trafficking.[31] To address this issue, NHS-activated biotin was conjugated to PEI in one reaction and PEG (5 kDa) succinimidyl propionate was conjugated to chicken avidin in a separate reaction.[32] The two copolymers were then coupled together by biotin–avidin interactions. When exposed to high salt conditions *in vitro*, the non-covalent interaction between biotin and avidin remained stable, suggesting the potential for prolonged systemic circulation as seen with covalently conjugated PEI-PEG copolymers. Further, addition of excess biotin reduced the degree of PEGylation and improved binding to the cell surface *in vitro*.[32]

To circumvent the charge reduction while incorporating PEG into PEI, we developed PEG-based homobifunctional derivatives for ionic cross-linking.[33] Ionic cross-linking was accomplished by using PEG-bis (phosphate) and resulted in nanoparticles of ~85–150 nm varying in accordance with the MW of PEG and degree of cross-linking. The PEI-PEG nanoparticles were 5–16-fold more efficient as transfecting agents compared with lipofectin and PEI itself (Figure 10.2).

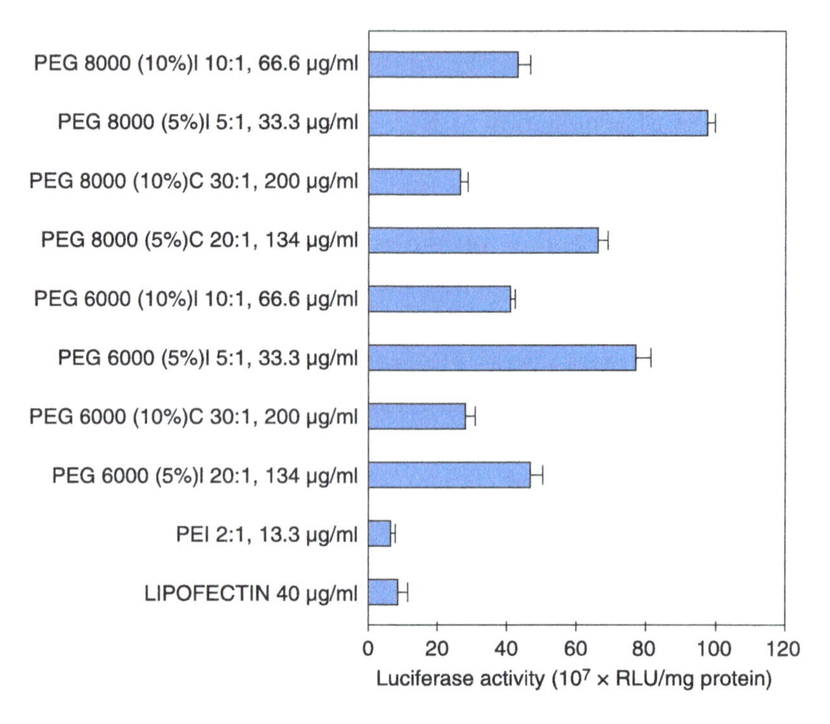

Figure 10.2 Comparison of transfection efficiency of various PEI-PEG nanoparticle complexes

COS-1 cells were incubated with PEI-PEG/DNA nanoparticle complexes at various weight ratios and incubated for 36 h. The luciferase reporter gene activity in the cell lysate was measured using a luminometer and the results are expressed in terms of relative light units/mg total cellular protein. The optimal transfection efficiency obtained by different nanoparticles is represented by the bar diagram. The absolute concentration of PEI-PEG nanoparticles complexed with DNA is also indicated. The assays were done in triplicate and the standard error is shown

Note: I, ionic cross-linked; C, covalent cross-linked
Source: adapted from Nimesh et al. 2010[33]

Published by Woodhead Publishing Limited, 2013

Further, the toxicity of PEI-PEG nanoparticles was found to be reduced considerably in comparison with PEI polymers.[33]

10.2.2 Hydrophobic modification of PEI

To improve transfection efficiency mediated through the balance between hydrophilicity and hydrophobicity, PEI was derivatized by introduction of alkyl linkage, attachment of hydrophobic ligands, etc. PEI was modified by linking cholesteryl chloroformate (PEI-Chol) to the secondary amino groups of BPEI (MW 1.8 and 10 kDa).[34] The mean particle size of the complexes was in the range 110–205 nm, except for the complexes prepared using PEI of 1.8 kDa, which had a mean particle size of 384 ± 300 nm. Further, the *in vitro* transfection of PEI-Chol/pCMS-EGFP complexes in Jurkat cells showed high levels of GFP expression with little toxicity.[34] In another study, N-alkylated LPEIs (MW 22 kDa) with varying alkyl chain lengths and extent of substitution were synthesized and investigated for *in vivo* transfection efficiency, specificity, and biodistribution.[35] N-Ethylation revealed improved *in vivo* efficacy of gene expression in the mouse lung by 26-fold relative to the parent polycation and more than quadrupled the ratio of expression in the lung to that in all other organs. N-propyl-PEI was the best performer in the liver and heart (581- and 3.5-fold enhancements) while N-octyl-PEI improved expression in the kidneys over the parent polymer 221-fold.[35] Further, the effects of the protonation properties of PEI were investigated by generation of PEI derivatives by acetylating varying fractions of the primary and secondary amines to form secondary and tertiary amides.[36] Acetylation of PEI decreased the physiological buffering capacity and decreased the zeta potential of polyplexes from 14 mV to 8–11 mV, with a two

Published by Woodhead Publishing Limited, 2013

to threefold increase in the polyplex diameter. Furthermore, acetylation had a negligible effect on cytotoxicity of PEI derivatives, while gene delivery efficacy increased by up to 21-fold compared with unmodified PEI, in both the presence and absence of serum.[36]

We partially derivatized the amino groups of PEI (750 kDa) by using three different acylating agents varying in carbon chain length from C-2 to C-4, to influence the proton sponge mechanism and hydrophobic–hydrophilic balance.[37] The acylated PEI-PEG nanoparticles were found to be in the range of 84–124 nm. The gene delivery efficacy in COS-1 cells was improved by 5–12-fold as compared with native PEI and the commercially available transfecting agent lipofectin, along with a significant reduction in cytotoxicity (Figure 10.3). Of all the systems prepared, nanoparticles with 30% acylation, using propionic anhydride, were found to be the most efficient at *in vitro* transfection.[37] In another study, the amino groups on the polymeric backbone of PEI were acylated using acetic or propionic anhydride.[38] Modified PEI showed reduced buffering capacity, and those polymers with buffering capacities greater than 50% and less than 80%, relative to PEI, showed higher transfection efficiencies than PEI. This study also showed PEI-propionic anhydride derivate to be the best transfecting system, with improved cell viability.[38]

10.2.3 Conjugation of targeting ligands

To improve their interaction with the cell surface, and ultimately their cellular uptake, PEIs can be modified with targeting ligands such as FA, the cell penetrating peptide TAT, RGD peptide, or galactose.[39–41] Active targeting strategies involve coupling of polyplexes with a ligand of

Published by Woodhead Publishing Limited, 2013

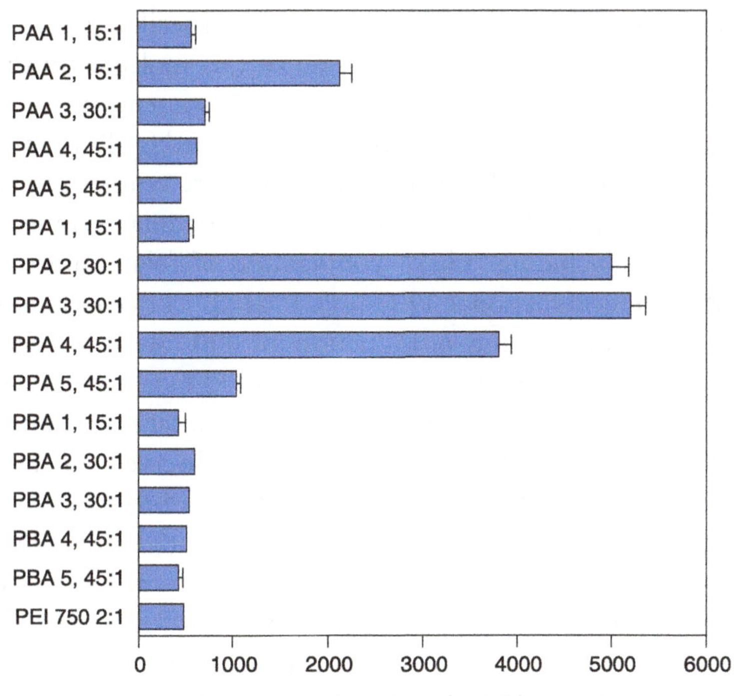

Figure 10.3 Comparison of transfection efficiency of various acylated PEI nanoparticle complexes

COS-1 cells were incubated with DNA-loaded acylated PEI nanoparticle complexes at various weight ratios for 4 h and the expression of GFP was monitored after 36 h. The fluorescent intensity of GFP fluorophore in the cell lysate was measured on a spectrofluorometer and the results are expressed in terms of arbitrary units/ mg total cellular protein. The results represent the mean of two independent experiments performed in triplicate. The data were recorded at optimal transfection efficiency for the respective acylated PEI nanoparticle:DNA ratio. Transfection efficiency with acylated PEI nanoparticles, DNA complexes for 750 kDa series

Source: adapted from Nimesh et al. 2010[33]

Published by Woodhead Publishing Limited, 2013

choice that is expected to interact with a specific target on the cell surface. An over-expression of this molecule in the target cell type is necessary if high uptake is intended.

Folate-conjugated ternary copolymers based on PEI-graft-PCL-block-PEG (PEI-g-PCL-b-PEG-FA) were synthesized as a targeted gene delivery system using a modular synthesis approach, including "click" conjugation of folate moieties with heterobifunctional PEG-b-PCL at the PEG terminus and subsequently the introduction of PEI by a Michael addition between folate-PEG-b-PCL and PEI via active PCL terminus.[42] An enhancement of cellular uptake of PEI-g-PCL-b-PEG-FA/pDNA polyplexes was observed in FRs overexpressing KB cells, which resulted in a 14-fold increase in transfection efficiency in comparison with unmodified PEI-g-PCL-b-PEG.[42] FA was conjugated to a backbone (named mPPS) consisting of a copolymer of methyl PEG-2000, PEI-600, and sebacoyl chloride for synthesis of novel polymer mPPS-FA.[39] Transfection studies revealed that mPPS-FA/DNA complexes yielded the highest GFP transfection efficiency in B16-F1O, U87, CHO-1, and HO-8910 cells, all of which highly express FRs, at an mPPS-FA/DNA ratio (w/w) of 15.

PEG-Tf-PEI conjugate yielded improved transfection efficiency in Jurkat cells and reduced cytotoxicity as compared with PEI complex.[43] This was attributed to a reduction in the membrane-damaging effect via shielding of the positive charge on the nanocomplex surface by PEG.[43] Further, to intensify the effects of the ligand on gene delivery, dual receptor-binding elements, Tf and transforming growth factor α (TGFα), were introduced into the PEI polyplex.[44] In A549, Tf and TGFα polyplex had higher uptake and transfection efficiency when compared with single Tf or TGFα introduced polyplex (Tf-polyplex and TGFα-polyplex), respectively, while no enhancement was observed in CHO-K1.[44]

Published by Woodhead Publishing Limited, 2013

PEG-PEI was coupled with a cyclic RGD peptide, a special ligand for integrin αvβ3 receptor, and employed as vector for pigment epithelial-derived factor (PEDF) gene therapy.[41] PEDF gene delivered by cyclic RGD-PEG-PEI apparently suppressed tumor growth, with a 67.4% reduction and decreased micro-vessel density in nude mice bearing SW620 human colorectal xenografts. HA-coated PEI-PBLG/DNA complexes were further modified by introducing RGD peptide (HA-RGD/PEI-PBLG/DNA) with grafting density of one RGD in every 1.9 HA repeating units.[45] The transfection efficiency of HA-RGD/PEI-PBLG/DNA was 9.7 times that of HA/PEI-PBLG/DNA for the RGD target binding affinity to the receptors on the HeLa cell surface. This was attributed to HA coating on PEI-PBLG/DNA reducing the electrostatic binding affinity to the cells, while the RGD tagging not only compensates for the reduced binding affinity for integrin on HeLa cells but also enhances the affinity for HA-RGD/PEI-PBLG/DNA.[45] In another study, PEI-PEG-based polyplexes containing MC1SP-peptide, a ligand specific for melanocortin receptor-1, demonstrated receptor-mediated transfection of Cloudman S91 (clone M-3) murine melanoma cells that was more efficient as compared with the non-targeted complexes.[46] Further, the targeted polyplexes carrying the HSVtk gene more efficiently inhibited melanoma tumor growth after ganciclovir administration and prolonged the lifespan of DBA/2 tumor-bearing mice compared with the non-targeted ones.

Key properties of the PEI-PEG-TAT peptide polyplex nanoparticles, including their behavior in cells, were investigated and compared with the transfection efficacy using 11 different cell lines.[40] A statistically significant positive correlation was found between transfection efficacy and the share of 50–75 nm fraction in the whole mixture of nanoparticles estimated with AFM. Also, variations in

Published by Woodhead Publishing Limited, 2013

PEG/PEI and N/P ratios made it possible to find their optimal combinations, which resulted in up to 100% transfection efficacy for several cell lines.[40] PEG (3.4 kDa)-g-PEI (25 kDa) modified with TAT polyplex mediated significantly higher transfection in lung epithelial cells of mice compared with TAT-PEI/pDNA polyplex, suggesting that covalent coupling of TAT to PEI via PEG led to high transfection.[47] These conjugates were able to transfect the epithelial cells of bronchi and alveoli. It has been reported that sugar-modified PEI can help *in vivo* gene delivery by enhancing specific interactions between the vector and target organs.

10.2.4 Coupling with other polymers

PEI(600 Da)-cyclodextrin (CD), prepared by linking low MW PEI and β-CD, was used to introduce the therapeutic gene TRAIL into mesenchymal stem cells (MSCs).[48] The particle size of PEI-CD measured at N/P 20 was around 180 ± 13 nm, with zeta potential around 29 mV. Further, cellular uptake studies revealed that the PEI-CD/DNA complex could escape from endosomes at a proper rate so that the transfection efficiency could be increased. Furthermore, TRAIL-MSCs reduced metastases in C57BL/6 mice by intravenous injection and did not show any signs of toxicity in the lung or liver of normal mice.[48] To reduce toxicity, Pun et al. synthesized CD-modified PEI derivatives (branched and linear).[49] While the cytotoxicity of both CD-BPEI and CD-LPEI was reduced with the increasing density of cyclodextrin, transfection efficiency of CD-BPEI was lower than that of the unmodified analogue despite higher cellular uptake. However, the transfection efficiency of both CD-BPEI and CD-LPEI was higher than for the unmodified analogues in the presence of chloroquine, suggesting that cyclodextrin conjugation hinders endosomal

Published by Woodhead Publishing Limited, 2013

escape. Later, adamantane-terminated PEG was synthesized, which could be non-covalently conjugated to CD-PEI. *In vivo* studies revealed no toxicity up to 120 μg DNA, a dose that is lethal for LPEI-DNA complexes. Biodistribution data showed highest accumulation in liver tumor, followed by the lungs, then the kidneys, with gene expression observed only in the liver.[49]

Shuai et al. incorporated PEG-PCL grafts onto the periphery of BPEI. These polyplexes showed reduced cytotoxicity and some of them exhibited improved transfection efficiency compared with BPEI (25 kDa).[50] The enhanced gene transfection efficiency of these PEG-PCL-PEI copolymers was later improved by creating inclusion complexes between the PEG-PCL grafts and α-CD.[51] PEI was also derivatized with PCL by Michael addition reaction between PCL diacrylate and low MW PEI to yield poly(ester amine)s (PEA).[52] Poly(ester amine)s/DNA complexes showed effective and stable DNA condensation with particle sizes below 200 nm. Poly(ester amine)s revealed much higher transfection efficiencies in three tested cell lines (293T, HepG2, HeLa cells) as compared with PEI (25 kDa), with PCL/PEI-1.2 (1.2 kDa) complex having transfection efficiency 15–25-fold higher. Further, poly(ester amine)s/DNA complexes transfected cells *in vivo* after aerosol administration more successfully than PEI (25 kDa).[52]

Chitosan and its derivatives have been studied as non-viral vectors due to biocompatibility, biodegradability, and low toxicity. To impart biodegradability, PEI-graft-chitosan was synthesized by cationic polymerization of aziridine in the presence of water-soluble chitosan (3.4 kDa).[53] PEI-graft-chitosan showed higher transfection efficiency and safety than PEI (25 kDa) in the different cells (HepG2, HeLa, and hepatocytes) due to the proton sponge effect of PEI in the polymer. Administration of the polymer/DNA complexes into the rat liver common bile duct showed 58-fold higher transfection efficiency in liver than PEI itself (25 kDa).[53] Later, PEI-graft-*N*-maleated chitosan was synthesized through grafting of low MW PEI (800 Da) to *N*-maleated chitosan by a Michael addition reaction.[54] The polymer revealed low cytotoxicity and good transfection efficiency in both 293T and HeLa cells, although high MW polymer showed higher cytotoxicity and lower transfection efficiency than low MW polymer.[54] In another study, PEI-graft-chitosan was prepared by grafting low MW PEI (600 Da) into the chitosan through a short PEG linker (440 Da) with terminal epoxide rings to decrease the inherent cytotoxicity of PEI-graft-chitosan.[55] The polymer showed higher cell viability than chitosan in 293T cells, and also mediated higher gene expression than chitosan in 293T cells.

10.3 Degradable PEI for gene delivery

Although PEI is highly efficient in transfecting mammalian cells, its non-degradability hampers its clinical application. The non-degradable cationic polymers could induce cytotoxic effects by destabilizing the cell membrane. Therefore, various chemical modifications that consist of biodegradable bonds are introduced into cationic polymers, in order to reduce the cytotoxicity as well as to control the release of DNA. The degradable, branched PEIs have a number of advantages over linear ones due to their high amine density. As the amine density in linear PEI is limited, it may not be enough to condense DNA efficiently. Branched PEIs are therefore more popular for the synthesis of degradable PEIs, although they need more stringent control over the reaction conditions due to the involvement of primary, secondary, and tertiary amines. Linear degradable PEI exhibits a short half-life, as

Published by Woodhead Publishing Limited, 2013

even a few cleavages can reduce chain length rapidly, with a quick drop in MW, whereas the branched PEI degrades slowly due to the lower water accessibility of the ester linkages in the branched structures.[56,57]

Though the introduction of biodegradable bonds could reduce cytotoxicity, the DNA delivery efficiency could also be decreased, probably due to reduction in the endosomal escape efficiency by the introduction of biodegradability, which causes a decrease in the amine density of the polymers. A biodegradable PEI derivative, LPEI-S, was synthesized for efficient and safe gene delivery.[58] The transfection efficiency was observed to increase as the amine density of LPEI-S increased. LPEI-S6 and LPEI-S8 show transfection efficiencies comparable to PEI with almost negligible cytotoxicity. Degradation studies of LPEI-S in HeLa cells revealed complete degradation within 3 h.[58] Further, the degradation rate of LPEI-S derivatives was controlled by varying the amount of the cross-linker bisepoxide.[59] The bisepoxide-PEI-S was readily degradable under reductive conditions (5 mM glutathione solution) and the degradation time was dependent on the degree of cross-linking. The transfection efficiency of bisepoxide-PEI-S was higher than that of LPEI-S; for instance, in NIH3T3s and HUVEC cells, the difference in transfection efficiency between LPEI-S and bisepoxide-PEI-S 5% was more than 1000-fold.[59]

In addition to disulfide linkages, PEI has been derivatized with acid-labile ester linkages for introduction of biodegradability. Biodegradable PEI-PEG conjugates were synthesized by reacting low MW PEI (600, 1200, 1800 Da) with PEG succinimidyl succinate (2000 Da) to form water-soluble polymers.[60] Neutralization of the complexes was achieved at charge ratios of copolymer/pSV-beta-gal plasmid from 0.8 to 1.0, with the mean particle size of the polyplexes ranging from 130 to 150 nm. *In vitro* transfection efficiency

Published by Woodhead Publishing Limited, 2013

of the synthesized copolymer was three-fold higher than that of the starting low MW PEI (1800 Da), while the cell viability was maintained at over 80%. To reduce the cytotoxicity, degradable PEIs with acid-labile imine linkers were synthesized with low MW PEI (MW 1.8 kDa) and glutadialdehyde.[61] The half-life of the acid-labile PEI was 1.1 h at pH 4.5 and 118 h at pH 7.4, suggesting that the acid-labile PEI may be rapidly degraded into non-toxic low MW PEI in the acidic endosome. *In vitro* transfection assays showed that the transfection efficiency of the acid-labile PEIs was comparable to that of PEI (25 kDa) and that they were far less toxic, due to the degradation of the acid-labile linkage.[61] In another study, biodegradable PEI-PEG copolymers were prepared by reaction of low MW PEI with PEG diacrylate as a cross-linker.[62] Particle sizes were observed to decrease with increasing N/P ratio and PEG MW, exhibiting a minimum value of 75 nm at an N/P ratio of 45 with PEI-PEG (MW 700 Da). Further, the transfection efficiency was also influenced by PEG MW and, in the case of PEI-PEG (MW 258 Da), the transfection efficiency was higher than that for PEI (25 kDa) in HepG2 and MG63, whereas it was lower in HeLa cells.[62]

Forrest et al. cross-linked low MW PEI (800 Da) with 1,3-butanediol (or 1,6-hexanediol) diacrylate as cross-linking agents to generate the ester-cross-linked polymer.[63] The acrylate groups reacted with both primary and secondary amines, resulting in highly branched, cross-linked, degradable PEI with a final MW of 14 kDa. The half-life of the cross-linked PEI synthesized from 1,3-butanediol was 4 h, due to the rapid hydrolysis of ester bonds in the polymeric structure at physiological conditions, to produce the diol linkers and amino acids.[63] Further, the degradable polymers exhibited similar size, structure, and DNA-binding properties to commercially available PEI (25 kDa), but

Published by Woodhead Publishing Limited, 2013

mediated 2–16-fold higher gene expression in MDA-MB-231 cells with lower cytotoxicity.[63]

10.4 Conclusions

The availability of LPEI and BPEI in a wide range of MWs with excellent transfection efficiencies and their implication in several studies to establish the optimal requirements are promising. From the available study database, it could be deduced that the optimal MW of PEI for complexation with DNA lies between 5 and 25 kDa, although high (800 kDa) and low (2 kDa) MW PEIs have also shown good transfection at their preferred N/P ratios. To evolve clinically relevant formulations, the design of a cationic PEI-based vector should be directed towards addressing the cytotoxicity concern, and efficient protection of DNA against serum or DNAse via efficient complexation with DNA.

10.5 References

1. Gebhart, C.L. and Kabanov, A.V. (2001) Evaluation of polyplexes as gene transfer agents. *J Control Release*, 73, 401–16.
2. Boussif, O., Lezoualc'h, F., Zanta, M.A., Mergny, M.D., Scherman, D., et al. (1995) A versatile vector for gene and oligonucleotide transfer into cells in culture and in vivo: polyethylenimine. *Proc Natl Acad Sci USA*, 92, 7297–301.
3. Neu, M., Fischer, D. and Kissel, T. (2005) Recent advances in rational gene transfer vector design based on poly(ethyleneimine) and its derivatives. *J Gene Med*, 7, 992–1009.

Published by Woodhead Publishing Limited, 2013

4. Jones, G.D., Langsjoen, A., Neumann, S.M.M.C. and Zomlefer, J. (1944) The polymerization of ethylenimine. *J Org Chem*, **09**, 125–47.
5. Brissault, B., Kichler, A., Guis, C., Leborgne, C., Danos, O., et al. (2003) Synthesis of linear polyethylenimine derivatives for DNA transfection. *Bioconjug Chem*, **14**, 581–7.
6. von Harpe, A., Petersen, H., Li, Y. and Kissel, T. (2000) Characterization of commercially available and synthesized polyethylenimines for gene delivery. *J Control Release*, **69**, 309–22.
7. Thomas, M., Lu, J.J., Ge, Q., Zhang, C., Chen, J., et al. (2005) Full deacylation of polyethylenimine dramatically boosts its gene delivery efficiency and specificity to mouse lung. *Proc Natl Acad Sci USA*, **102**, 5679–84.
8. Thomas, M. and Klibanov, A.M. (2002) Enhancing polyethylenimine's delivery of plasmid DNA into mammalian cells. *Proc Natl Acad Sci USA*, **99**, 14640–5.
9. Akinc, A., Thomas, M., Klibanov, A.M. and Langer, R. (2005) Exploring polyethylenimine-mediated DNA transfection and the proton sponge hypothesis. *J Gene Medicine*, **7**, 657–63.
10. Goula, D., Benoist, C., Mantero, S., Merlo, G., Levi, G., et al. (1998) Polyethylenimine-based intravenous delivery of transgenes to mouse lung. *Gene Ther*, **5**, 1291–5.
11. Abdallah, B., Hassan, A., Benoist, C., Goula, D., Behr, J.P., et al. (1996) A powerful nonviral vector for in vivo gene transfer into the adult mammalian brain: polyethylenimine. *Hum Gene Ther*, **7**, 1947–54.
12. Erbacher, P., Bettinger, T., Brion, E., Coll, J.L., Plank, C., et al. (2004) Genuine DNA/polyethylenimine (PEI) complexes improve transfection properties and cell survival. *J Drug Target*, **12**, 223–36.

Published by Woodhead Publishing Limited, 2013

13. Kunath, K., von Harpe, A., Fischer, D., Petersen, H., Bickel, U., et al. (2003) Low-molecular-weight polyethylenimine as a non-viral vector for DNA delivery: comparison of physicochemical properties, transfection efficiency and in vivo distribution with high-molecular-weight polyethylenimine. *J Control Release*, **89**, 113–25.
14. Fischer, D., Bieber, T., Li, Y., Elsasser, H.P. and Kissel, T. (1999) A novel non-viral vector for DNA delivery based on low molecular weight, branched polyethylenimine: effect of molecular weight on transfection efficiency and cytotoxicity. *Pharm Res*, **16**, 1273–9.
15. Godbey, W.T., Wu, K.K. and Mikos, A.G. (1999) Size matters: molecular weight affects the efficiency of poly(ethylenimine) as a gene delivery vehicle. *J Biomed Mater Res*, **45**, 268–75.
16. Ogris, M., Brunner, S., Schuller, S., Kircheis, R. and Wagner, E. (1999) PEGylated DNA/transferrin-PEI complexes: reduced interaction with blood components, extended circulation in blood and potential for systemic gene delivery. *Gene Ther*, **6**, 595–605.
17. Baker, A., Saltik, M., Lehrmann, H., Killisch, I., Mautner, V., et al. (1997) Polyethylenimine (PEI) is a simple, inexpensive and effective reagent for condensing and linking plasmid DNA to adenovirus for gene delivery. *Gene Ther*, **4**, 773–82.
18. Meunier-Durmort, C., Grimal, H., Sachs, L.M., Demeneix, B.A. and Forest, C. (1997) Adenovirus enhancement of polyethylenimine-mediated transfer of regulated genes in differentiated cells. *Gene Ther*, **4**, 808–14.
19. Kleemann, E., Jekel, N., Dailey, L.A., Roesler, S., Fink, L., et al. (2009) Enhanced gene expression and reduced toxicity in mice using polyplexes of low-molecular-weight

Published by Woodhead Publishing Limited, 2013

poly(ethyleneimine) for pulmonary gene delivery. *J Drug Target*, **17**, 638–51.

20. Wightman, L., Kircheis, R., Rossler, V., Carotta, S., Ruzicka, R., et al. (2001) Different behavior of branched and linear polyethylenimine for gene delivery in vitro and in vivo. *J Gene Med*, **3**, 362–72.
21. Goula, D., Remy, J.S., Erbacher, P., Wasowicz, M., Levi, G., et al. (1998) Size, diffusibility and transfection performance of linear PEI/DNA complexes in the mouse central nervous system. *Gene Ther*, **5**, 712–17.
22. Sung, S.J., Min, S.H., Cho, K.Y., Lee, S., Min, Y.J., et al. (2003) Effect of polyethylene glycol on gene delivery of polyethylenimine. *Biol Pharm Bull*, **26**, 492–500.
23. Petersen, H., Fechner, P.M., Martin, A.L., Kunath, K., Stolnik, S., et al. (2002) Polyethylenimine-graft-poly(ethylene glycol) copolymers: influence of copolymer block structure on DNA complexation and biological activities as gene delivery system. *Bioconjug Chem*, **13**, 845–54.
24. Dash, P.R., Read, M.L., Barrett, L.B., Wolfert, M.A. and Seymour, L.W. (1999) Factors affecting blood clearance and in vivo distribution of polyelectrolyte complexes for gene delivery. *Gene Ther*, **6**, 643–50.
25. Kunath, K., Merdan, T., Hegener, O., Haberlein, H. and Kissel, T. (2003) Integrin targeting using RGD-PEI conjugates for in vitro gene transfer. *J Gene Med*, **5**, 588–99.
26. Sagara, K. and Kim, S.W. (2002) A new synthesis of galactose-poly(ethylene glycol)-polyethylenimine for gene delivery to hepatocytes. *J Control Release*, **79**, 271–81.
27. Kichler, A., Chillon, M., Leborgne, C., Danos, O. and Frisch, B. (2002) Intranasal gene delivery with a polyethylenimine-PEG conjugate. *J Control Release*, **81**, 379–88.

Published by Woodhead Publishing Limited, 2013

28. Nguyen, H.K., Lemieux, P., Vinogradov, S.V., Gebhart, C.L., Guerin, N., et al. (2000) Evaluation of polyether-polyethylenimine graft copolymers as gene transfer agents. *Gene Ther*, **7**, 126–38.
29. Fitzsimmons, R.E.B. and Uluda, H. (2012) Specific effects of PEGylation on gene delivery efficacy of polyethylenimine: Interplay between PEG substitution and N/P ratio. *Acta Biomater*, **8**, 3941–55.
30. Merdan, T., Kunath, K., Petersen, H., Bakowsky, U., Voigt, K.H., et al. (2005) PEGylation of poly(ethylene imine) affects stability of complexes with plasmid DNA under in vivo conditions in a dose-dependent manner after intravenous injection into mice. *Bioconjug Chem*, **16**, 785–92.
31. Godbey, W.T., Wu, K.K. and Mikos, A.G. (1999) Tracking the intracellular path of poly(ethylenimine)/DNA complexes for gene delivery. *Proc Natl Acad Sci USA*, **96**, 5177–81.
32. Xiong, M.P., Forrest, M.L., Karls, A.L. and Kwon, G.S. (2007) Biotin-triggered release of poly(ethylene glycol)-avidin from biotinylated polyethylenimine enhances in vitro gene expression. *Bioconjug Chem*, **18**, 746–53.
33. Nimesh, S., Goyal, A., Pawar, V., Jayaraman, S., Kumar, P., et al. (2006) Polyethylenimine nanoparticles as efficient transfecting agents for mammalian cells. *J Control Release*, **110**, 457–68.
34. Wang, D.A., Narang, A.S., Kotb, M., Gaber, A.O., Miller, D.D., et al. (2002) Novel branched poly(ethylenimine)-cholesterol water-soluble lipopolymers for gene delivery. *Biomacromolecules*, **3**, 1197–207.
35. Fortune, J.A., Novobrantseva, T.I. and Klibanov, A.M. (2011) Highly effective gene transfection in vivo by alkylated polyethylenimine. *Journal of Drug Delivery*, **2011**, Article ID 204058, doi:10.1155/2011/204058.

Published by Woodhead Publishing Limited, 2013

36. Forrest, M.L., Meister, G.E., Koerber, J.T. and Pack, D.W. (2004) Partial acetylation of polyethylenimine enhances in vitro gene delivery. *Pharm Res*, **21**, 365–71.
37. Nimesh, S., Aggarwal, A., Kumar, P., Singh, Y., Gupta, K.C., et al. (2007) Influence of acyl chain length on transfection mediated by acylated PEI nanoparticles. *Int J Pharm*, **337**, 265–74.
38. Aravindan, L., Bicknell, K.A., Brooks, G., Khutoryanskiy, V.V. and Williams, A.C. (2009) Effect of acyl chain length on transfection efficiency and toxicity of polyethylenimine. *Int J Pharm*, **378**, 201–10.
39. Xu, Z., Jin, J., Siu, L.K.S., Yao, H., Sze, J., et al. (2012) Folic acid conjugated mPEG-PEI600 as an efficient non-viral vector for targeted nucleic acid delivery. *Int J Pharm*, **426**, 182–92.
40. Ulasov, A.V., Khramtsov, Y.V., Trusov, G.A., Rosenkranz, A.A., Sverdlov, E.D., et al. (2011) Properties of PEI-based polyplex nanoparticles that correlate with their transfection efficacy. *Mol Ther*, **19**, 103–12.
41. Li, L., Yang, J., Wang, W.-W., Yao, Y.-C., Fang, S.-H., et al. (2012) Pigment epithelium-derived factor gene loaded in cRGD–PEG–PEI suppresses colorectal cancer growth by targeting endothelial cells. *Int J Pharm*, **438**, 1–10.
42. Liu, L., Zheng, M., Renette, T. and Kissel, T. (2012) Modular synthesis of folate conjugated ternary copolymers: polyethylenimine-graft-polycaprolactone-block-poly(ethylene glycol)-folate for targeted gene delivery. *Bioconjug Chem*, **23**, 1211–20.
43. Lee, K.M., Lee, Y.B. and Oh, I.J. (2011) Evaluation of PEG-transferrin-PEI nanocomplex as a gene delivery agent. *J Nanosci Nanotechnol*, **11**, 7078–81.
44. Kakimoto, S., Moriyama, T., Tanabe, T., Shinkai, S. and Nagasaki, T. (2007) Dual-ligand effect of transferrin

Published by Woodhead Publishing Limited, 2013

and transforming growth factor alpha on polyethylenimine-mediated gene delivery. *J Control Release*, **120**, 242–9.

45. Tian, H., Lin, L., Chen, J., Chen, X., Park, T.G. and Maruyama, A. (2011) RGD targeting hyaluronic acid coating system for PEI-PBLG polycation gene carriers. *J Control Release*, **155**, 47–53.
46. Durymanov, M.O., Beletkaia, E.A., Ulasov, A.V., Khramtsov, Y.V., Trusov, G.A., et al. (2012) Subcellular trafficking and transfection efficacy of polyethylenimine–polyethylene glycol polyplex nanoparticles with a ligand to melanocortin receptor-1. *J Control Release*, **163**, 211–19.
47. Kleemann, E., Neu, M., Jekel, N., Fink, L., Schmehl, T., et al. (2005) Nano-carriers for DNA delivery to the lung based upon a TAT-derived peptide covalently coupled to PEG-PEI. *J Control Release*, **109**, 299–316.
48. Hu, Y.-L., Huang, B., Zhang, T.-Y., Miao, P.-H., Tang, G.-P., et al. (2012) Mesenchymal stem cells as a novel carrier for targeted delivery of gene in cancer therapy based on nonviral transfection. *Mol Pharm*, **9**, 2698–709.
49. Pun, S.H., Bellocq, N.C., Liu, A., Jensen, G., Machemer, T., et al. (2004) Cyclodextrin-modified polyethylenimine polymers for gene delivery. *Bioconjug Chem*, **15**, 831–40.
50. Shuai, X., Merdan, T., Unger, F., Wittmar, M. and Kissel, T. (2003) Novel biodegradable ternary copolymers hy-PEI-g-PCL-b-PEG: synthesis, characterization, and potential as efficient nonviral gene delivery Vectors. *Macromolecules*, **36**, 5751–9.
51. Shuai, X., Merdan, T., Unger, F. and Kissel, T. (2005) Supramolecular gene delivery vectors showing enhanced

Published by Woodhead Publishing Limited, 2013

transgene expression and good biocompatibility. *Bioconjug Chem*, **16**, 322–9.

52. Arote, R., Kim, T.H., Kim, Y.K., Hwang, S.K., Jiang, H.L., et al. (2007) A biodegradable poly(ester amine) based on polycaprolactone and polyethylenimine as a gene carrier. *Biomaterials*, **28**, 735–44.
53. Wong, K., Sun, G., Zhang, X., Dai, H., Liu, Y., et al. (2006) PEI-g-chitosan, a novel gene delivery system with transfection efficiency comparable to polyethylenimine in vitro and after liver administration in vivo. *Bioconjug Chem*, **17**, 152–8.
54. Lu, B., Xu, X.D., Zhang, X.Z., Cheng, S.X. and Zhuo, R.X. (2008) Low molecular weight polyethylenimine grafted N-maleated chitosan for gene delivery: properties and in vitro transfection studies. *Biomacromolecules*, **9**, 2594–600.
55. Lou, Y.L., Peng, Y.S., Chen, B.H., Wang, L.F. and Leong, K.W. (2009) Poly(ethylene imine)-g-chitosan using EX-810 as a spacer for nonviral gene delivery vectors. *J Biomed Mater Res A*, **88**, 1058–68.
56. Wu, D., Liu, Y., Jiang, X., He, C., Goh, S.H., et al. (2006) Hyperbranched poly(amino ester)s with different terminal amine groups for DNA delivery. *Biomacromolecules*, **7**, 1879–83.
57. Anderson, D.G., Akinc, A., Hossain, N. and Langer, R. (2005) Structure/property studies of polymeric gene delivery using a library of poly(beta-amino esters). *Mol Ther*, **11**, 426–34.
58. Lee, Y., Mo, H., Koo, H., Park, J.Y., Cho, M.Y., et al. (2007) Visualization of the degradation of a disulfide polymer, linear poly(ethylenimine sulfide), for gene delivery. *Bioconjug Chem*, **18**, 13–18.

Published by Woodhead Publishing Limited, 2013

59. Koo, H., Jin, G.W., Kang, H., Lee, Y., Nam, K., et al. (2010) Biodegradable branched poly(ethylenimine sulfide) for gene delivery. *Biomaterials*, **31**, 988–97.
60. Ahn, C.H., Chae, S.Y., Bae, Y.H. and Kim, S.W. (2002) Biodegradable poly(ethylenimine) for plasmid DNA delivery. *J Control Release*, **80**, 273–82.
61. Kim, Y.H., Park, J.H., Lee, M., Kim, Y.-H., Park, T.G., et al. (2005) Polyethylenimine with acid-labile linkages as a biodegradable gene carrier. *J Control Release*, **103**, 209–19.
62. Park, M.R., Han, K.O., Han, I.K., Cho, M.H., Nah, J.W., et al. (2005) Degradable polyethylenimine-alt-poly(ethylene glycol) copolymers as novel gene carriers. *J Control Release*, **105**, 367–80.
63. Forrest, M.L., Koerber, J.T. and Pack, D.W. (2003) A degradable polyethylenimine derivative with low toxicity for highly efficient gene delivery. *Bioconjug Chem*, **14**, 934–40.

Published by Woodhead Publishing Limited, 2013

11

Atelocollagen

DOI: 10.1533/9781908818645.225

Abstract: Atelocollagen was the first naturally occurring biomaterial with potential application as gene delivery vector and is prepared by pepsin treatment from type I collagen of calf dermis. Atelocollagen/DNA complexes can be fabricated into beads, sponge, membrane, minipellet, etc. without use of heat or without using any organic solvent. Further, high concentration of atelocollagen allows the complex to be stable for prolonged times, which is advantageous for a sustained release carrier. On the contrary, low concentration of atelocollagen results in formation of complex particles with diameter in size 100–300 nm, which is considered adequate for systemic applications. Furthermore, treatment with atelocollagen did not alter expression level of toxicity related genes, suggesting it to be practically non-toxic and potential candidate for gene vector. Atelocollagen has been successfully employed in several *in vitro* and *in vivo* gene delivery studies. The present chapter accounts for the studies done to establish the gene delivery applications of atelocollagen.

Key words: atelocollagen, minipellet, nanosize, antitumor, tetrahydrobiopterin.

Published by Woodhead Publishing Limited, 2013

11.1 Introduction

Atelocollagen was the first naturally occurring biomaterial with potential application as gene delivery vector.[1] It is prepared by pepsin treatment from type I collagen of calf dermis.[2,3] The N- and C-terminals of the collagen molecules possess an amino acid sequence called telopeptide, which determines the antigenicity of collagen. However, pepsin treatment gets rid of telopeptides in atelocollagen, which eliminates its immunogenicity and empowers clinical applications. At low temperature atelocollagen exists as a liquid, which favors complexation with nucleic acids. Since the surface of atelocollagen molecules is positively charged, the molecules can bond electrostatically with negatively charged nucleic acid molecules. Upon implantation in the body, it exhibits plasticity, by which it becomes fibrous and then solid due to body temperature. This solidification allows entrapment of DNA in the mesh structure of the matrix, resulting in protection from immunological reaction and enzymatic attack (Figure 11.1). Moreover, atelocollagen/ DNA complexes can be fabricated into beads, sponge, membrane, minipellet, etc. without use of heat and without using any organic solvent, which is a major cause of deactivation of gene vectors.

Atelocollagen was used for the preparation of minipellet containing pDNA, pH adjusted to neutral, to maintain stability of pDNA. To prevent degradation of DNA chains due to lyophilization, glucose is added as a lyoprotective additive (Figure 11.2). Further, the size of the complex particles can be modulated by varying the ratio of nucleic acid to atelocollagen. Furthermore, a high concentration of atelocollagen allows the complex to be stable for a prolonged time, which is advantageous for a sustained release carrier. On the other hand, a low concentration of atelocollagen

Published by Woodhead Publishing Limited, 2013

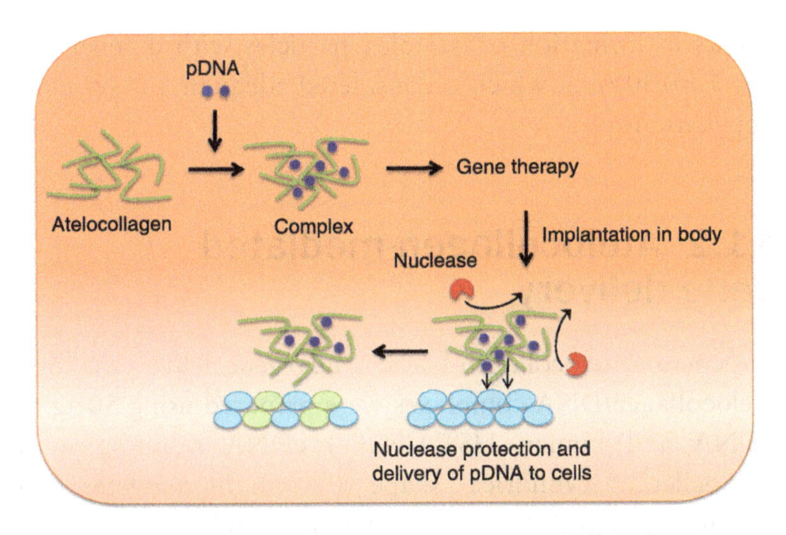

Figure 11.1 Mechanism of atelocollagen-mediated gene delivery

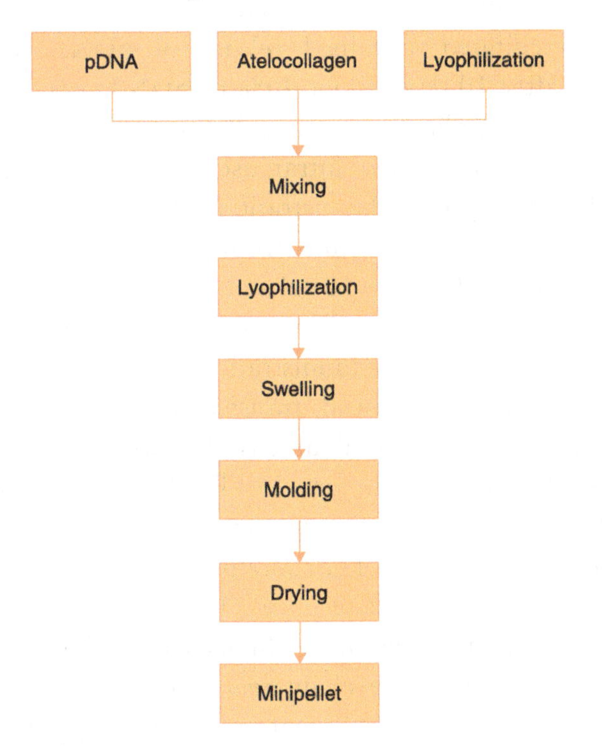

Figure 11.2 Preparation of minipellet containing pDNA

Published by Woodhead Publishing Limited, 2013

results in formation of complex particles with diameter of size 100–300 nm, which is considered adequate for systemic applications.

11.2 Atelocollagen-mediated gene delivery

To generate atelocollagen-based *in vivo* gene delivery vectors, atelocollagen/DNA complexes were prepared using 50 μg of pDNA and human HST-1/FGF-4 cDNA resulting in a minipellet of cylindrical shape (0.6 mm in diameter and 10 mm in length).[1,4] The optimization studies involved *in vitro* pDNA release from the minipellet with or without glucose. A minipellet comprising 30% (weight/weight) glucose facilitated sustained release of pDNA into the medium; after 10 days about 40% of pDNA was released.[1] For *in vivo* application, minipellet containing 50 μg of pCAHST-1 was injected intramuscularly into 7-week-old normal ICR mice. Administration of minipellet led to increase in the platelet counts from a pre-injection level of $98 \pm 4 \times 10^4$/μl to $145 \pm 9 \times 10^4$/μl and $168 \pm 12 \times 10^4$/μl at days 10 and 28, respectively, remaining elevated for up to 60 days. On the other hand, administration of 50 μg of naked pCAHST-1 pDNA showed transiently increased platelet counts of $175 \pm 12 \times 10^4$/μl at day 10, which returned to normal ($98 \pm 4 \times 10^4$/μl) by day 25.[1] Further, the serum levels of HST-1/FGF-4 were significantly increased upon administration of minipellet containing 50 μg of pCAHST-1 pDNA, with the highest level of 112 ± 22 pg/ml on day 30 after injection and remaining as high as 50–100 pg/ml for up to 60 days. Polymerase chain reaction (PCR) analysis of released pDNA revealed that pCAHST-1-specific PCR product first appeared in the sera of mice 6 h after injection and remained present for 40 days, while in controls PCR products were barely detectable after 21 days.[1] It was deduced that the *in vivo* administration of the minipellet resulted in serum protein levels of HST-1/FGF-4, which produced a 200% increase in the platelet count, sufficient to inhibit experimentally induced thrombocytopenia in mice.[5]

Due to its cationic charge and binding properties, the ability of atelocollagen to transfer cDNA expression vectors, AS-ODNs, and adenovirus vectors into cells on a microplate was investigated.[6] In this study, atelocollagen/genetic material complexes as nanosized particles were pre-coated on a micro-well plate onto which the cells were then seeded. Atelocollagen (80 μg/ml) condensed pDNA (7 kb) (100 μg/ml) to form complexes in the nanoparticle form with a diameter of ~100–200 nm, which yielded maximal transfection efficacy.[6] In one example, HEK 293 cells were added to a plate coated with atelocollagen/GFP reporter gene nanoparticles. The fluorescence intensity due to expression of GFP in the cells was observed to occur in a dose-dependent manner that remained up to 52 days post-transfection. Moreover, the spotted and air-dried nanoparticles on the plate retained their transfection ability for more than 1 month when stored at room temperature. Further, to demonstrate the utility of atelocollagen in ODN delivery, it was employed to deliver AS-ODNs that specifically inhibited the growth of human testicular tumor cells, NEC8, which showed high expression levels of HST-1/FGF-4 mRNA.[6] The results revealed significant growth-inhibiting effects on cells with AS-ODNs against HST-1/FGF-4, while mock ODNs had no effects. Also, the growth inhibition by the complex of AS-ODNs against human HST-1/FGF-4 occurred in an ODN-dose-dependent manner.[6] Furthermore, atelocollagen was involved in delivery of an adenovirus vector carrying a GFP gene. The embryonic stem cells plated

Published by Woodhead Publishing Limited, 2013

onto the atelocollagen/adenovirus-prefixed plate expressed GFP in a virus-titer-dependent manner, while the adenovirus solution alone prefixed to the plate as a control did not lead to GFP gene expression in cells.[6]

Over-expression of the fibroblast growth factor HST-1/FGF-4 gene is associated with growth and malignancy in human testicular germ cell tumors. To evaluate the therapeutic relevance of HST-1/FGF-4 AS-ODNs directed against human HST-1/FGF-4 mRNA, they were delivered by complexation with atelocollagen and analyzed for anti-tumor activity.[7] Among the four AS-ODNs specific for the HST-1/FGF-4 gene, A4196T (58.8% inhibition) and A5773V (22.2% inhibition) showed efficient reduction of proliferation of NEC8 cells *in vitro* at 2 μM concentration. The AS-ODNs also inhibited the cell proliferation of other male germ cell tumors which produce a high amount of HST-1/FGF-4, such as NEC14 and NCCIT cells.[7] Further, the AS-ODNs significantly suppressed the secretion of HST-1/FGF-4 in NEC8 cells treated with 5 μM ODNs for 72 h. Assessment 6 h after intratesticular administration of atelocollagen/ODNs revealed the presence of significant amounts of ODNs in tumors as compared with ODNs injected alone. After 20 days of ODNs administration, the tumor volume in mice treated with atelocollagen/AS-ODNs was 68, 78, or 76% smaller than that in mice treated with AS-ODNs alone, control sense ODNs alone, or atelocollagen alone, respectively.[7] The HST-1/FGF-4 protein level in tumors was significantly inhibited in tumors ($P < 0.001$) after 3 days of administration of atelocollagen/AS-ODNs. Furthermore, atelocollagen-mediated transfer of AS-ODNs significantly inhibited tumor metastasis in the para-aortic lymph nodes and lungs.[7]

To study the potential application of atelocollagen as an ODN delivery carrier *in vivo*, the activity of formulated

Published by Woodhead Publishing Limited, 2013

AS-ODN targeted against the ICAM-1 mRNA was investigated in an allergic dermatitis mice model.[8] AS-ODN was administered to the animals as a single intravenous injection of formulation containing 100 μg/ml of AS-ODN in 0.05% atelocollagen solution. Antisense activity was determined by measurement of ear thickness, histopathology, and immunohistochemistry, 24 h after the initiation of the dermatitis. A single dose of atelocollagen/AS-ODN at a dosage of 0.6 mg/kg showed almost complete inhibition of the ear swelling and was comparable to non-inflamed tissue in the unsensitized group.[8] The immunohistochemical analysis showed reduced ICAM-1 expression at the challenged site in animals treated with the atelocollagen/AS-ODN. Also, the antisense activity was found to increase according to the concentration of atelocollagen in the formulation.[8] Further, the effect mediated by the atelocollagen/AS-ODN was more than 50 times greater than that provided by ODN infusion, although the level of ODN formulated with atelocollagen dropped below that of the 24 h infusion group within 30 min. Moreover, atelocollagen/AS-ODN suppressed inflammatory progression by treatment at 8 h after the ear challenge, when inflammation had already commenced at the challenged site. Furthermore, antisense activity was noted even when the atelocollagen/AS-ODN was injected 3 days before the initiation of inflammation.[8]

To treat tetrahydrobiopterin (BH_4) deficiency by over-expression of its biosynthetic enzyme, GTP cyclohydrolase I (GTP-CH1), a new gene transfer method was developed by insertion into a pCAGGS vector, followed by complexation of pDNA with atelocollagen.[9] The BH_4 content and GTP-CH1 activity in Zucker fatty rats were 50–55% less than in Zucker lean rats. *In vivo* administration of atelocollagen/GTP-CH1 complexes in Zucker fatty rats significantly improved aortic BH_4 content to more than 80%

Published by Woodhead Publishing Limited, 2013

of the level in Zucker lean rats. Further, atelocollagen/GTP-CH1 complexes stimulated O_2^- production, along with reduction in endothelial nitric oxide synthase (eNOS) activity, and endothelial function in insulin-resistant Zucker fatty rats was improved by a plasmid DNA injection to a level similar to that in Zucker lean rats.[9]

To evaluate the effect of ornithine decarboxylase (ODC) AS-ODNs on *in vitro* cell growth of gastrointestinal cancer (MKN 45 and COLO201) and RD and *in vivo* in mice, atelocollagen/AS-ODN complexes were employed.[10] Significant suppression in MKN45, COLO201, and rhabdomyosarcoma (RD) cell growth, by 83.8%, 86.5%, and 95.6%, was obtained with atelocollagen/AS-ODN. Also, atelocollagen/AS-ODN remarkably suppressed tumor growth over a 35–42-day period. Further, the tumor suppression was found to be dependent on route of administration; for instance, intramuscular and intraperitoneal injection remarkably suppressed RD tumors, while intraperitoneal administration completely suppressed growth of MKN45 and COLO201 tumors at 35 days.[10] Furthermore, intratumoral administration of intraperitoneal atelocollagen/AS-ODN significantly suppressed ODC activities in MKN45, COLO201, and RD tumors: 87.8%, 64.9%, and 71.3%, respectively.[10]

Atelocollagen has been used as an effective drug delivery technology to sustain the release of anti-tumor proteins and to enhance the anti-tumor activity of ODNs in *in vivo* models. Hence, it is imperative to investigate the toxicity of atelocollagen for successful clinical applications. Towards this end, whole genome expression profiling in mouse liver was done after systemic administration of atelocollagen or cationic liposome DOTP/cholesterol (LP) and compared with hepatotoxicity.[11] Microarray analysis revealed that systemic LP administration significantly elevated the expression level of toxicity-related genes, i.e. lipocalein 2,

Published by Woodhead Publishing Limited, 2013

cyclin dependent kinase inhibitor 1A, serum amyloid A isoforms, chemokine ligands, and garnzyme B. However, treatment with atelocollagen did not alter any of these, suggesting that atelocollagen is practically non-toxic and can be a potential biomaterial for gene and drug delivery.[11] In a recent study, the effects of atelocollagen on endothelial paracellular barrier function were investigated.[12] Atelocollagen/ODN complexes significantly increased the permeability of two types of endothelial cells, and the change was dependent on the molecular size, structure, and concentration of the ODNs and atelocollagen used. Further, immune-histochemical studies suggested that complexes influenced the cellular skeleton and intercellular structure, but without affecting the expression of adherent junction or tight junction proteins, and these changes were induced through p38 MAP kinase signaling.[12]

11.3 Conclusions

Atelocollagen has been successfully employed in several *in vitro* and *in vivo* gene delivery studies. Further, treatment with atelocollagen did not alter the expression level of toxicity-related genes, suggesting that it is practically non-toxic and a potential candidate for a gene vector. However, the mechanical aspects of atelocollagen-mediated gene delivery still need to be elucidated in order to develop vectors for clinical applications.

11.4 References

1. Ochiya, T., Takahama, Y., Nagahara, S., Sumita, Y., Hisada, A., et al. (1999) New delivery system for plasmid

DNA in vivo using atelocollagen as a carrier material: the minipellet. *Nat Med*, **5**, 707–10.

2. Ochiya, T., Nagahara, S., Sano, A., Itoh, H. and Terada, M. (2001) Biomaterials for gene delivery: atelocollagen-mediated controlled release of molecular medicines. *Curr Gene Ther*, **1**, 31–52.
3. Sano, A., Maeda, M., Nagahara, S., Ochiya, T., Honma, K., et al. (2003) Atelocollagen for protein and gene delivery. *Adv Drug Deliv Rev*, **55**, 1651–77.
4. Fujioka, K., Takada, Y., Sato, S. and Miyata, T. (1995) Novel delivery system for proteins using collagen as a carrier material: the minipellet. *J Control Release*, **33**, 307–15.
5. Konishi, H., Ochiya, T., Sakamoto, H., Tsukamoto, M., Saito, I., et al. (1995) Effective prevention of thrombocytopenia in mice using adenovirus-mediated transfer of HST-1 (FGF-4) gene. *J Clin Invest*, **96**, 1125–30.
6. Honma, K., Ochiya, T., Nagahara, S., Sano, A., Yamamoto, H., et al. (2001) Atelocollagen-based gene transfer in cells allows high-throughput screening of gene functions. *Biochem Biophys Res Commun*, **289**, 1075–81.
7. Hirai, K., Sasaki, H., Sakamoto, H., Takeshita, F., Asano, K., et al. (2003) Antisense oligodeoxynucleotide against HST-1/FGF-4 suppresses tumorigenicity of an orthotopic model for human germ cell tumor in nude mice. *J Gene Med*, **5**, 951–7.
8. Hanai, K., Kurokawa, T., Minakuchi, Y., Maeda, M., Nagahara, S., et al. (2004) Potential of atelocollagen-mediated systemic antisense therapeutics for inflammatory disease. *Hum Gene Ther*, **15**, 263–72.
9. Shinozaki, K., Nishio, Y., Yoshida, Y., Koya, D., Ayajiki, K., et al. (2005) Supplement of tetrahydrobiopterin by a

Published by Woodhead Publishing Limited, 2013

gene transfer of GTP cyclohydrolase I cDNA improves vascular dysfunction in insulin-resistant rats. *J Cardiovasc Pharmacol*, **46**, 505–12.

10. Nakazawa, K., Nemoto, T., Hata, T., Seyama, Y., Nagahara, S., et al. (2007) Single-injection ornithine decarboxylase-directed antisense therapy using atelocollagen to suppress human cancer growth. *Cancer*, **109**, 993–1002.
11. Ogawa, S., Onodera, J., Honda, R. and Fujimoto, I. (2011) Influence of systemic administration of atelocollagen on mouse livers: an ideal biomaterial for systemic drug delivery. *J Toxicol Sci*, **36**, 751–62.
12. Hanai, K., Kojima, T., Ota, M., Onodera, J. and Sawada, N. (2012) Effects of atelocollagen formulation containing oligonucleotide on endothelial permeability. *Journal Drug Deliv*, **2012**, 245835.

Published by Woodhead Publishing Limited, 2013

12

Protamine nanoparticles

DOI: 10.1533/9781908818645.237

Abstract: Protamine is a small, arginine-rich, nuclear protein found in the sperm and is purified from the mature testes of salmon fish, with MW ~4000–6000 Da. Protamine sulfate have been shown to be a safer and more appropriate alternative to poly-L-lysine for condensation, as well as the delivery of plasmid DNA to the nucleus. Also, protamine salts have been often used in combination with liposomal preparations to deliver plasmid DNA into cells. These ternary complexes were designed to mimic viral vectors containing condensed DNA, which is surrounded by a lipid bilayer and termed as LPDs. Further, to design novel anionic ternary nanoparticles for gene delivery, protamine/DNA complexes (150–200 nm) were coated with anionic solid lipid nanoparticles (~20 nm). The present chapter evidences the studies done towards development of protamine as gene delivery vector.

Key words: protamines, protamine sulfate, transferrin, oligonucleotides, apolipoprotein, LPDs.

12.1 Introduction

Protamine is a small, arginine-rich, nuclear protein that replaces histones in the late haploid phase of spermatogenesis.

Published by Woodhead Publishing Limited, 2013

More than 67% of the amino acid composition in protamine is arginine. It is a naturally occurring protein found only in sperm and is purified from the mature testes of fish, usually salmon, with MW ~4000–6000 Da. In sperm, protamine binds to DNA, facilitating formation of a compact structure followed by delivery of DNA to the nucleus of the egg after fertilization. This unique functionality overcomes a major obstacle in gene therapy by non-viral vectors, i.e. the efficient delivery of DNA from the cytoplasm into the nucleus. Protamine sulfate has been shown to be a safer and more appropriate alternative to poly-L-lysine for condensation, as well as the delivery of plasmid DNA to the nucleus. Also, protamine salts have often been used in combination with liposomal preparations to deliver plasmid DNA into cells.

12.2 Protamine nanoparticles for gene delivery

In one of the initial studies, one to three protamine molecules were linked to one transferrin molecule to create a DNA delivery system similar to the transferrin-PLL conjugates. The transferrin-protamine/DNA (luciferase encoding pDNA pRSVL) conjugates were recognized and taken up by receptor-mediated endocytosis in avian erythroblasts. This delivery system showed low cytotoxic side effects in this cell line.[1] ODNs mixed with protamine free base solution spontaneously formed nanoparticles with a diameter of 150–170 nm, which increased up to ~300–330 nm after 3 days of incubation.[2] The nanoparticles prepared at mass ratios of 0.5:1 (protamine/ODN) and 1:1 were negatively charged while those at 2:1 were positively charged. The nanoparticles were highly stable on exposure to fetal calf serum (FCS) and cell culture medium.[2] Further, nanoparticles carrying an antisense ODN

Published by Woodhead Publishing Limited, 2013

directed towards the proto-oncogene c-myc were readily taken up by human promonocytic leukemia cells (U 937) and showed significant cellular growth inhibition (~50% growth inhibition by 2.5:1 mass ratio nanoparticles).[2] Another study reported formation of nanoparticles in the range of 90–150 nm by complexation of protamine/ODNs.[3] For nanoparticle formation a minimal chain length of nine nucleotides and a mass ratio of 0.5:1 were required. The surface charge and the number of nanoparticles were dependent on the mass ratio. Uptake studies in African green monkey kidney cells (Vero cells) revealed mass ratio-dependent cellular uptake of nanoparticles, and internalized ODNs were seen in the cytoplasm and nucleus of the cells. The nanoparticles protected ODNs against nuclease degradation and were almost non-toxic.[3] This nuclease stability of ODNs was also observed in nanoparticles prepared with various protamine salts and treated with rat small intestine homogenates.[4] Protamine sulfate and protamine chloride significantly improved the nuclease stability of the ODNs. This study was pursued to prove the potential utility of protamine/ODN nanoparticles for gastro-intestinal administration.[4]

The interaction of protamine salts and ODN was investigated to determine the physico-chemical characteristics of the resulting complex systems and to analyze the influence of permeation enhancers (sodium chenodeoxycholate and sodium caprate) on the dissociation of the complexes.[5] Zeta potential data confirmed the conductometric equivalence points and explained the good physical stability of charged complexes when compared with neutral complexes (+/–25 mV for protamine sulfate complexes). Moreover, incorporation of sodium chenodeoxycholate promoted complex dissociation, while sodium caprate inhibited dissociation.[5] For efficient transfection of HIV-1 target cells, protamine was used to complex ODN and phosphorothioate (PTO) analogues to

Published by Woodhead Publishing Limited, 2013

form nanoparticles with diameters of about 180 nm and surface charges in the range of −18 to +30 mV.[6] The uptake of these nanoparticles was significantly improved as compared with naked oligonucleotides. Protamine/ODN nanoparticles showed release of the antisense compound leading to specific inhibition of tat-mediated HIV-1 transactivation. On the other hand, protamine/PTO complexes were stable over 72 h, and failed to release AS-PTO.[6]

To harness the similarity between low MW protamine fragments (LMWPs) and HIV-TAT protein transduction peptide, protamine was enzymatically digested with thermolysin to yield LMWPs and investigated for gene delivery potential.[7] The size and zeta potential of the LMWP (1.9 kDa)/pDNA complexes were 120 nm and +30 mV, respectively, which were quite suitable for cellular uptake. After complexation LMWPs appeared to effectively protect pDNA against DNase I attack. Confocal microscopic analysis of the two peptides confirmed the cytoplasmic (in 15 min) and nuclear localization (in 1 h) of the peptides with a lower sign of adhesion to the cell membranes.[7] Also, the transfection assay of pSV-β-galactosidase into the 293T cells employing LMWP/DNA nanoparticles showed maximal efficacy at a N/P ratio of 10, which was comparable to that of TAT/DNA complexes and significantly higher than that of the PEI/DNA complexes, while exhibiting a markedly reduced cytotoxicity compared with the pDNA/PEI complex.[7] To decipher the interaction of DNA with peptides and determine the critical factors that control expression following gene delivery, protamine was selectively digested enzymatically to produce LMWPs of various lengths and amino acid compositions.[8] The lower MW fractions of LMWPs (0.6–2.4 kDa) condensed DNA to form particles of size 1000–3000 nm, while the higher MW fraction (2.8–3.0 kDa) formed nanoparticles of size ~142 nm, as compared with protamine, which formed nanoparticles of size

Published by Woodhead Publishing Limited, 2013

~278 nm. Further, all the LMWP peptides showed lower DNA binding strength as compared with protamine. Also, the LMWP peptide-mediated *in vitro* gene delivery showed prolonged (up to 12 days) gene expression in HeLa and CHO cells.[8]

The advantages associated with protamine/ODN nanoparticles include: (1) simple and rapid preparation by self-assembly; (2) non-antigenic and possess very low toxicity; (3) the particles are comparatively stable in water and, at least under favorable conditions, in cell culture medium; (4) high uptake by a variety of cells, as compared with the naked ODNs; and (5) after cellular uptake, they dissociate readily, resulting in rapid release of ODNs. Although protamine/ODN nanoparticles showed promising results, they also suffer from disadvantages: (i) tendency to aggregate in aqueous solutions under isotonic conditions; (ii) poor escape of nanoparticles entrapped inside endosomes. To overcome these problems, HSA, a non-toxic, biodegradable macromolecule, was introduced as a protective colloid.[9] Protamine sulphate and HSA complexed to form particles of size ~10–14 nm that increased to 230–320 nm after addition of ODNs. The surface charge of the particles ranged from about −12 to +60 mV depending on the protamine concentration and the ionic conditions.[9] Uptake studies in mouse fibroblast cells revealed a 12-fold increased cellular uptake of ODNs using nanoparticles in comparison to free oligonucleotides, while 100% of the cells were transfected.[10] These nanoparticles showed an enhanced release of ODN into the cytoplasm compared with the protamine/ODN nanoparticles. Moreover, the nanoparticles showed antisense effect of about 35% in a functional assay as well as on the protein level (western blot) along with very low cytotoxicity during a 24h application. However, substitution of the protamine free base by protamine sulfate led to a drastic

Published by Woodhead Publishing Limited, 2013

reduction in size, with nanoparticles of only around 40 nm in diameter with otherwise unchanged properties.[11] These small-sized nanoparticles may be advantageous when dealing with cells which show size-dependent particle uptake.

A comparative study was done with two commercially available liposomes (DOTAP, lipofectin), one artificial virus capsoid (polyoma VP1), two cationic acrylate nanoparticles, and two protamine-based nanoparticles.[12] The protamine-based nanoparticles were prepared by complexation of protamine with ODN, either alone or later followed by HSA coating. The HSA-coated nanoparticles showed more diffuse ODN distribution throughout the cytoplasm as compared with uncoated ones. Further, the HSA-coated nanoparticles showed significantly improved antisense effect in comparison to uncoated nanoparticles.[12]

Drug delivery to the brain is severely restricted by formation of tight junctions between adjacent brain capillary endothelial cells (BCECs). In order to develop a delivery system to BCEC cells, protamine/ODN nanoparticles were coated with apolipoprotein A-I (apoA-I), as apolipoproteins were suggested to enhance cellular uptake of nanoparticles.[13] Adsorption of apoA-I on the surface of nanoparticles resulted in significantly improved uptake and transcytosis properties as compared with uncoated proticles. Further, aβoA-I coating enhanced nanoparticle delivery to astrocytes almost twofold in an *in vitro* model of the BBB.[13]

In order to achieve target-specific cellular binding, a cationic derivative of dextran using protamine (Dex-P) was developed.[14] Evaluation of particle size showed the condensation of Dex-P/ctDNA (calf thymus DNA) nanoplexes to a diameter range of 6–7 nm. Also, Dex-P showed considerable buffering capacity in the pH range from 7 to 6. Complexation of DNA with Dex-P provided protection from the effect of DNase as compared with naked

Published by Woodhead Publishing Limited, 2013

DNA digested with DNase I.[14] A hemolysis study performed with Dex-P showed no detectable red blood cell (RBC) membrane disruptive activity, thus proving it to be non-hemolytic. Further, RBC aggregation studies with Dex-P resulted in absence of aggregation. Dex-P did not activate C3 to a greater extent than protamine, whereas PEI was found to strongly activate the complement system.[14] Cell viability studies after exposure of Dex-P to L929 cells suggested that reduction in cytotoxicity was due to modification of dextran with protamine. Furthermore, the transfection potential of Dex-P/pGL3 nanoparticles in HepG2 cells was comparable to that with BPEI (25 kDa). The transfection efficacy was found to correlate with the localization of the plasmid within the nucleus and with increase in incubation time. Entry of the plasmid into the nucleus was also increased.[14]

Recently, superoxide dismutase 1 (SOD1) was conjugated with LMWP and the effect of LMWP-SOD1 conjugates on hydrogen peroxide-induced cellular senescence and osteoblastic differentiation was investigated.[15] LMWP-SOD1 significantly increased senescence-associated β-galactosidase activity with enlargement and flattening of human dental pulp stem cell (DPSC) morphology. Further, LMWP-SOD1 abolished activation of the cell cycle regulator proteins, p53 and $p21^{Cip1}$, induced by hydrogen peroxide. Additionally, LMWP-SOD1 reversed the inhibition of osteoblastic differentiation and downregulation of osteogenic gene markers induced by hydrogen peroxide.[15]

12.3 Liposome/protamine/ DNA complexes

Liposomes have become one of the most widely studied non-viral vectors since their introduction as potential gene

Published by Woodhead Publishing Limited, 2013

carriers.[16] They comprise a group of positively charged lipids at physiological pH that interact with the negatively charged DNA through electrostatic attractions. Initial studies with liposomes appeared promising; however, their therapeutic applications were limited due to the formation of large particles (0.6–1 μm) at optimal transfection efficiencies. To circumvent this shortcoming and further improve transfection efficacy, the DNA was condensed with a polycationic peptide such as protamine before binding to the cationic carrier.[17] These ternary complexes were designed to mimic viral vectors containing condensed DNA surrounded by a lipid bilayer and termed LPDs (Figure 12.1). The pre-condensation

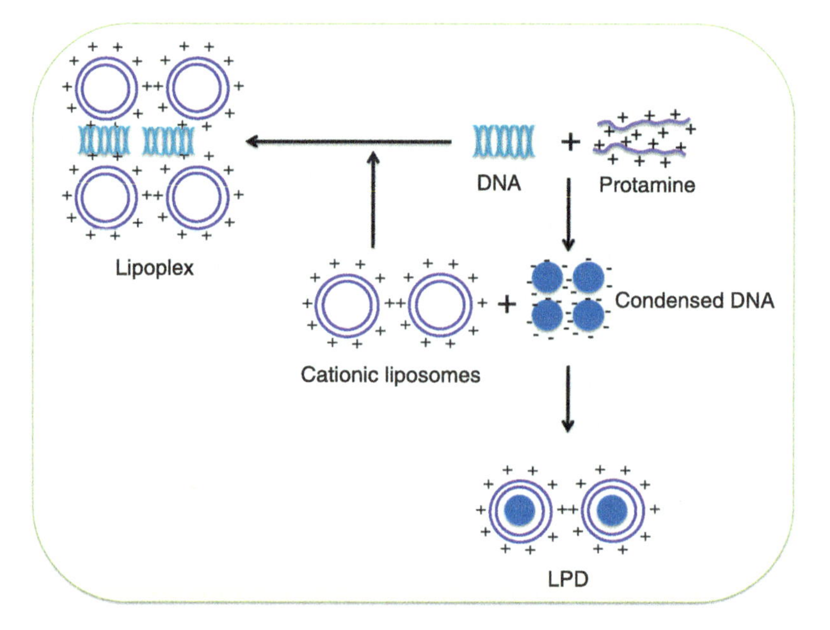

Figure 12.1 Schematic representation of comparison between LPD and lipoplex particles. DNA is condensed and neutralized with protamine before entrapment into liposomes to form the LPD complex, while in lipoplexes DNA is sandwiched between liposomal particles

Published by Woodhead Publishing Limited, 2013

of DNA with polycationic peptide resulted in a reduction of the particle size and an increased stability against nucleases.[17] Also, LPDs formed at a higher initial ratio of liposome to DNA had higher lipid content and were highly active in transfection; the activity was about three to ninefold more than the corresponding complex before purification.[17] Sorgi et al. described the ability of several different protamine species to enhance the transfection activity of cationic lipids.[18] The structural differences between the protamines were accounted for by the differences among various salt forms of protamine. The appearance of lysine residues within the protamine molecule correlated well with a reduction in binding affinity to plasmid DNA, as well as an observed loss in transfection activity. In another study, You et al. observed improvement in transfection efficiency by adding protamine to plasmid DNA solution before the formation of DNA-lipid vesicle complexes.[19] Both free-base protamine and protamine sulfate revealed better transfection efficiency and expression level, but the optimal amount of the two protamines was different. The increment in transfection efficiency and gene expression level was at most 20-fold compared with dimethyldioctadecyl ammonium bromide lipid vesicles. Protamines were thought to protect DNA from degradation by DNAse and promote delivery into the nucleus.[19]

To develop stable gene delivery systems suitable for long-term storage, lyophilization of cationic lipidic vector LPD nanoparticles of size ~100 nm were investigated as a model.[20] With the use of a sufficient amount of appropriate lyoprotectant during lyophilization, both particle size and transfection efficiency could be preserved. Evaluation of a series of monosaccharides and disaccharides, including dextrose, galactose, mannose, lactose, maltose, sucrose, and trehalose, for their lyoprotective effect revealed disaccharides as better protectors than monosaccharides.[20] No significant

Published by Woodhead Publishing Limited, 2013

difference was found with the use of different freezing protocols for lyophilization; however, for freeze-thawing, fast freezing caused less aggregation. Further, lyophilization of LPD with 10% sucrose allowed storage at room temperature without significant change in particle size or loss of transfection efficiency.[20] Another study investigated LPD complexes comprising protamine that gave rise to multi-lamellar lipid-coated nanoparticles of size ~100 nm, smaller and more homogeneous compared with complexes of lipid/pDNA alone of size ~145 nm, and whose structure was unaffected by jet nebulization.[21] Further, the LPD complexes provided remarkably higher protection of pDNA against the shearing forces encountered in a jet nebulizer as compared with lipid/pDNA complexes prepared even at very high mass (w:w) ratios, i.e. 12:1. To circumvent poor transfection efficiency in the presence of serum, protamine was covalently bound to stearic acid, followed by incorporation into lipid vesicles of dimethyldioctadecyl ammonium bromide (DDAB) to make stable complexes between DNA and DDAB lipid vesicles.[22] The optimal ratio of DDAB to protamine (5:1) gave about 50% transfection efficiency in COS-7 cells, which remained unaltered even in the presence of 10% serum. Further, protamine-modified DDAB lipid vesicles also enhanced virus transduction efficiency in the presence of serum using a replication-defective retroviral vector. Furthermore, the vesicles allowed efficient gene transfer for avian embryos *in vivo*.[22]

The intracellular fate of ODN delivered by LPD in HepG2 cells was investigated by condensation of fluorescent labeled 68-mer ODN with protamine and coated with three different liposomal formulations.[23] LPDs effectively transported protamine-complexed ODNs into the nucleus, while both the negatively charged and neutral control liposomes, as well as lipid/ODN complexes in the absence of protamine, did

Published by Woodhead Publishing Limited, 2013

not induce any nuclear uptake of ODN after 21 h of incubation.[23] Later, a comparative study was done between a protamine/DNA complex coated with a lipid envelope made of cationic DOTAP, i.e. LPD, and a DOTAP/DNA lipoplex based on the same lipid composition.[24] The transfection efficacy was evaluated in four different cell lines, i.e. CHO (Chinese hamster ovary cells), HEK293 (human embryonic kidney cells), NIH 3T3 (mouse embryonal cells), and A17 (murine cancer cells). The transfection efficiency of LPD was found to be from three to 20 times higher than that of DOTAP/DNA lipoplexes. Physico-chemical properties were found to control the ability of nanocarriers to release DNA upon interaction with cellular membranes. LPD complexes easily release their DNA payload, while lipoplexes remain largely intact and accumulate at the nucleus in the cell.[24]

The transfection efficacy of LPD nanoparticles was reported to be ~39-fold more than liposome/DNA complexes.[25] The order of protamine addition significantly influenced the transfection of LPD nanoparticles; higher efficacy was obtained with LPD nanoparticles prepared with pre-mixed protamine/DNA complexes. Intracellular tracking studies evidenced the presence of ODNs in nucleus when transfected with LPD, while they were only distributed in cytoplasm with liposome/DNA complexes.[25] To evaluate the anti-tumor effects of p27 gene therapy on pulmonary metastatic tumors by a non-viral gene delivery strategy, LPD complexes were prepared.[26] The size and zeta potential of LPD nanoparticles were ~100 nm and +50 mV, which remained unaltered after lyophilization and storage. The *in vivo* gene expression data suggest that, after intravenous injection, LPD has targeting transduction ability with respect to lung tissues, to which it showed much higher transfection efficiency in comparison with other organs. Further, combined treatment with LPD and the chemotherapy agent cisplatin resulted in better therapeutic effects compared with cisplatin treatment or LPD treatment alone.[26]

A novel stearyl protamine was synthesized by the reaction of natural protamine with an NHS ester of stearic acid for gene delivery.[27] The higher stearyl group content derivative self-assembled into complexes with a size of 110 nm in water and had relatively higher *in vitro* gene transfection efficiency than protamine and lower stearyl group content derivatives. The high transfection efficiency was attributed to increased ability to destabilize the cell membrane, facilitating the entry of DNA into the cell via endocytosis and efficient dissociation of complexes with the introduction of hydrophobic stearyl groups into protamine.[27]

Lipid nanoparticles with improved cellular uptake capacity were utilized to prepare lipid nanoparticles/protamine/DNA ternary nanoparticles for gene delivery.[28] The anionic lipid nanoparticles consisting of monostearin and different contents of oleic acid had an average size range of 23.6 to 71.3 nm and a zeta potential about −30 mV. The average size of the protamine/DNA complex at 2:1 mass ratio was 128.5 nm, with +19.4 mV zeta potential. Lipid nanoparticles/ protamine/DNA ternary nanoparticles had an average size ranging from 192.7 to 260.6 nm and were negatively charged. Cellular uptake studies suggest that the uptake ability of lipid nanoparticles was enhanced by increasing the oleic acid content. Lipid nanoparticles with 20 wt% oleic acid/ protamine/DNA ternary nanoparticles (w/w/w, 65:6:3) showed maximal transfection efficacy that remained unaffected even in the presence of serum.[28]

Pulmonary endothelium is vital for normal pulmonary physiology and its dysfunction is involved in a number of pulmonary diseases. To develop a vector targeting pulmonary endothelium of lung via the vascular route, LPD was developed that was composed of DOTAP liposomes, protamine, and

Published by Woodhead Publishing Limited, 2013

ODN.[29] Intracellular tracking studies suggested efficient accumulation of ODN in the alveolar capillary region with targeted delivery to the pulmonary endothelial cells. The delivery efficiency was found to be dependent on lipid composition and the charge ratio between lipid and ODN. Furthermore, this formulation was associated with minimal pro-inflammatory cytokine response and other hematologic toxicities when the ODN lacked a potent unmethylated CpG motif. Also, pre-treatment of mice with LPD containing an ODN against intercellular adhesion molecule-1 (ICAM-1) significantly decreased ICAM-1 expression in the lung following lipopolysaccharide (LPS) challenge.[29]

The incorporation of pegylated lipid into lipid-protamine-DNA (LPD-PEG) lipopolyplexes causes a decrease of their *in vitro* transfection activity. To restore transfection efficacy and achieve tumor-specific delivery of LPD, RGD peptide was incorporated into the LPD.[30] Conjugation of RGD into LPD complexes resulted in a fivefold and 15-fold increase in binding and uptake, respectively, which led to a 100-fold enhancement of transfection activity. Moreover, this transfection enhancement was specific to cells expressing appropriate integrin receptors (MDA-MB-231 cells).[30]

The liver performs several functions vital for well-being, which makes this organ very attractive for gene therapy. However, the proportion of administered macromolecules internalized by hepatocytes depends on their particle size and biochemical characteristics. A liver-targeting lipidic vector was developed by pre-complexation of pDNA with different amounts of protamine, followed by the addition of cationic DOTAP/Chol liposome-asialofetuin (AF) complexes.[31] Use of AF amounts below 1 μg led to the formation of protamine-AF-lipoplexes of ~300 nm with a positive zeta potential. The resulting protamine-AF-lipoplexes remarkably enhanced the levels of gene expression in cultured HepG2 cells and in

Published by Woodhead Publishing Limited, 2013

the liver upon intravenous administration. Further, lipoplexes comprising the optimal amount of AF (1 μg/μg DNA) showed a 16-fold higher transfection activity in HepG2 cells than non-targeted (plain) complexes. Luciferase gene expression in the liver of mice after systemic administration of protamine-AF-lipoplexes via the tail vein increased by 12-fold as compared with plain complexes.[31] In order to develop a liver-targeted biocompatible gene delivery system, protamine/DNA condensates were complexed with PEG derivative having both carboxylic acid and lactose side chains (Lac-PEG-C).[32] The size of the nanoparticles was 180–200 nm, with high stability even in high ionic strength solutions. Lac-PEG-C coating onto protamine/DNA complexes reduced their zeta potential and inhibited albumin-induced aggregation. Further, the nanoparticles did not induce coagulation of RBCs, and possessed low cytotoxicity. Addition of Lac-PEG-C to protamine/DNA complexes prior to incubation with HepG2 cells markedly enhanced gene expression, and the weight ratio of 1.5:1:8 resulted in a 56-fold higher expression of luciferase than that without Lac-PEG-C.[32] LPD complexes composed of galactosylated cationic liposomes, protamine sulfate, and pDNA significantly improved the levels of gene expression in cultured hepatoma cells HepG2 and SMMC-7721 as compared with non-galactosylated LPD.[33] The transfection efficacy was observed to be influenced by the length of the spacer between the anchor and galactose residues for recognition by the asialoglycoprotein receptor.

12.4 Protamine conjugation to other ligands

Photochemical internalization (PCI) is a photodynamic therapy-based approach for improving the delivery of

Published by Woodhead Publishing Limited, 2013

macromolecules and genes into the cell cytosol.[34] The utility of PCI for the delivery of the GFP reporter gene on the same plasmid as a tumor suppressor gene (PTEN) was investigated in monolayers of U251 human glioma cells and U87 glioma spheroids employing protamine/DNA (GFP/PTEN or GFP) complexes. Protamine/GFP complexes were practically non-toxic to the glioma cells, but were highly inefficient at gene transfection if used alone. PCI treatment induced a five to tenfold increase in transfection efficacy as compared with non-treated controls. Further, PCI-protamine/PTEN transfection of either U251 monolayers or U87 spheroids significantly inhibited their growth but had no effect on MCF-7 cells containing a wild-type PTEN gene.[34]

To design novel anionic ternary nanoparticles for gene delivery, protamine/DNA complexes (150–200 nm) were coated with anionic solid lipid nanoparticles (SLNs) (~20 nm).[35] The SLNs were synthesized by a modified film dispersion–ultrasonication method, and adsorbed onto the protamine/DNA complexes via electrostatic interaction. The resultant ternary nanoparticles were observed to be uniform in size (257.7 ± 10.6 nm) and the surface charge inversion was from 19.28 ± 1.14 mV to −17.16 ± 1.92 mV.[35] Further, circular dichroism spectra analysis confirmed coating of SLNs onto protamine/DNA complexes without alteration of their structure. The ternary nanoparticles provided enhanced protection of DNA from nuclease degradation and higher cell viability of A549 cells compared with protamine/DNA complexes and lipofectamine. Furthermore, the transfection efficiency of the ternary nanoparticles was comparable to that of the commercially available transfection reagent lipofectamine and much higher than that of naked DNA and the protamine/DNA complexes.[35]

Published by Woodhead Publishing Limited, 2013

The transfection capacity of a new multi-component system based on dextran (Dex), protamine (Prot), and solid lipid nanoparticles (SLN) was evaluated *in vitro* and *in vivo* after intravenous administration to mice.[36,37] The DNA-SLN vector and the Dex-Prot-DNA-SLN vector showed sizes ~250–285 nm, and zeta potential about ~+30 to +35 mV. Dex-Prot-DNA-SLN nanoparticles induced a sixfold increase in the transfection of SLNs in retinal cells (ARPE 19 cells) due to the presence of protamine, which improved the DNA protection capacity, and a shift in the internalization mechanism from caveolae/raft-mediated to clathrin-mediated endocytosis.[37] However, Dex-Prot-DNA-SLN nanoparticles resulted in an almost complete inhibition of transfection in HEK-293 cells. Further evaluation of transfection efficiency in four different cell lines showed a higher transfection efficacy in the cells with a high ratio of activity of clathrin/caveolae-mediated endocytosis.[36] The complex prepared without dextran and protamine (DNA-SLN) was more efficient in cells with a high ratio of activity of caveolae/clathrin-mediated endocytosis. Further, the Dex-Prot-DNA-SLN vector showed no agglutination of erythrocytes, an effect probably attributed to the presence of dextran. Finally, intravenous administration of Dex-Prot-DNA-SLN vector to BALB/c mice induced GFP expression in liver, spleen, and lungs, and the protein expression was maintained for at least 7 days.[36]

Quaternary complexes with a condensed core of plasmid DNA, protamine, fish sperm DNA, and a shell of stearic acid grafted chitosan oligosaccharide (CSO-SA) were prepared.[38] The CSO-SA self-assembled in aqueous solution to form 25 nm micelles, which were smaller than CSO-SA micelles and CSO-SA micelles/plasmid DNA binary complexes. The transfection efficiencies of quaternary complexes on HEK293 and MCF-7 cells increased with incubation time, and were

Published by Woodhead Publishing Limited, 2013

significantly higher than that of CSO-SA micelles/plasmid DNA binary complexes. The optimal transfection efficiency of quaternary complexes on HEK293 and MCF-7 cells after 96h was 23.82% and 41.43%, respectively, which was comparable to Lipofectamine 2000. Further, gene expression analysis showed that the optimal ratio of plasmid DNA:fish sperm DNA:protamine:CSO-SA was 1:1:5:5.[38]

To evaluate the effect of the amount of protamine used as the transfection promoter in SLN-mediated gene delivery, three protamine-SLN samples (Pro25, Pro100, and Pro200) were prepared by adding increasing amounts of protamine.[39] The size of SLN was ~230nm, and only Pro200 showed a few particle aggregates. The SLN samples (Pro100 and Pro200) exhibited good pDNA complexation as compared with Pro25. Among these, only Pro100, having an intermediate amount of protamine, appeared able to promote pDNA cell transfer, especially in a neuronal cell line (Na1300). The amount of protamine as the transfection promoter in SLN was observed to affect not only the gene delivery ability of SLN but also its capacity to transfer genes efficiently to specific cell types.[39]

12.5 Conclusions

Although protamine has been successfully employed in numerous *in vitro* gene delivery studies, a clear understanding of the factors influencing the transfection efficiency still needs to be developed. Incorporation of protamine into liposomes led to a dramatic increase in the transfection efficiency along with a decrease in the cytotoxicity of the resultant LPD. However, there still remains a huge gap between *in vitro* studies with protamine-based nanoparticles for gene delivery and clinical applications.

Published by Woodhead Publishing Limited, 2013

12.6 References

1. Wagner, E., Zenke, M., Cotten, M., Beug, H. and Birnstiel, M.L. (1990) Transferrin-polycation conjugates as carriers for DNA uptake into cells. *Proc Natl Acad Sci U S A*, **87**, 3410–14.
2. Junghans, M., Kreuter, J. and Zimmer, A. (2000) Antisense delivery using protamine-oligonucleotide particles. *Nucleic Acids Res*, **28**, E45.
3. Junghans, M., Kreuter, J. and Zimmer, A. (2001) Phosphodiester and phosphorothioate oligonucleotide condensation and preparation of antisense nanoparticles. *Biochim Biophys Acta*, **1544**, 177–88.
4. González Ferreiro, M, Crooke, R.M., Tillman, L., Hardee, G. and Bodmeier, R. (2003) Stability of polycationic complexes of an antisense oligonucleotide in rat small intestine homogenates. *Eur J Pharm Biopharm*, **55**, 19–26.
5. González Ferreiro, M., Tillman, L., Hardee, G. and Bodmeier, R. (2001) Characterization of complexes of an antisense oligonucleotide with protamine and poly-L-lysine salts. *J Control*, **73**, 381–90.
6. Dinauer, N., Lochmann, D., Demirhan, I., Bouazzaoui, A., Zimmer, A., et al. (2004) Intracellular tracking of protamine/antisense oligonucleotide nanoparticles and their inhibitory effect on HIV-1 transactivation. *J Control Release*, **96**, 497–507.
7. Park, Y.J., Liang, J.F., Ko, K.S., Kim, S.W. and Yang, V.C. (2003) Low molecular weight protamine as an efficient and nontoxic gene carrier: in vitro study. *J Gene Med*, **5**, 700–11.
8. Kharidia, R., Friedman, K.A. and Liang, J.F. (2008) Improved gene expression using low molecular weight peptides produced from protamine sulfate. *Biochemistry (Mosc)*, **73**, 1162–8.

Published by Woodhead Publishing Limited, 2013

9. Lochmann, D., Weyermann, J., Georgens, C., Prassl, R. and Zimmer, A. (2005) Albumin-protamine-oligonucleotide nanoparticles as a new antisense delivery system. Part 1: physicochemical characterization. *Eur J Pharm Biopharm*, **59**, 419–29.
10. Weyermann, J., Lochmann, D., Georgens, C. and Zimmer, A. (2005) Albumin-protamine-oligonucleotide-nanoparticles as a new antisense delivery system. Part 2: cellular uptake and effect. *Eur J Pharm Biopharm*, **59**, 431–8.
11. Mayer, G., Vogel, V., Weyermann, J., Lochmann, D., van den Broek, J.A., et al. (2005) Oligonucleotide-protamine-albumin nanoparticles: Protamine sulfate causes drastic size reduction. *J Control Release*, **106**, 181–7.
12. Weyermann, J., Lochmann, D. and Zimmer, A. (2004) Comparison of antisense oligonucleotide drug delivery systems. *J Control Release*, **100**, 411–23.
13. Kratzer, I., Wernig, K., Panzenboeck, U., Bernhart, E., Reicher, H., et al. (2007) Apolipoprotein A-I coating of protamine-oligonucleotide nanoparticles increases particle uptake and transcytosis in an in vitro model of the blood-brain barrier. *J Control Release*, **117**, 301–11.
14. Thomas, J.J., Rekha, M.R. and Sharma, C.P. (2010) Dextran-protamine polycation: an efficient nonviral and haemocompatible gene delivery system. *Colloids Surf B Biointerfaces*, **81**, 195–205.
15. Choi, Y.J., Lee, J.Y., Chung, C.P. and Park, Y.J. (2012) Cell-penetrating superoxide dismutase attenuates oxidative stress-induced senescence by regulating the p53-p21(Cip1) pathway and restores osteoblastic differentiation in human dental pulp stem cells. *Int J Nanomedicine*, **7**, 5091–106.

Published by Woodhead Publishing Limited, 2013

16. Felgner, P.L., Gadek, T.R., Holm, M., Roman, R., Chan, H.W., et al. (1987) Lipofection: a highly efficient, lipid-mediated DNA-transfection procedure. *Proc Natl Acad Sci U S A*, **84**, 7413–17.
17. Gao, X. and Huang, L. (1996) Potentiation of cationic liposome-mediated gene delivery by polycations. *Biochemistry*, **35**, 1027–36.
18. Sorgi, F.L., Bhattacharya, S. and Huang, L. (1997) Protamine sulfate enhances lipid-mediated gene transfer. *Gene Ther*, **4**, 961–8.
19. You, J., Kamihira, M. and Iijima, S. (1999) Enhancement of transfection efficiency by protamine in DDAB lipid vesicle-mediated gene transfer. *J Biochem*, **125**, 1160–7.
20. Li, B., Li, S., Tan, Y., Stolz, D.B., Watkins, S.C., et al. (2000) Lyophilization of cationic lipid–protamine–DNA (LPD) complexes. *J Pharm Sci*, **89**, 355–64.
21. Birchall, J.C., Kellaway, I.W. and Gumbleton, M. (2000) Physical stability and in-vitro gene expression efficiency of nebulised lipid–peptide–DNA complexes. *Int J Pharm*, **197**, 221–31.
22. Mizuarai, S., Ono, K., You, J., Kamihira, M. and Iijima, S. (2001) Protamine-modified DDAB lipid vesicles promote gene transfer in the presence of serum. *J Biochem*, **129**, 125–32.
23. Welz, C., Neuhuber, W., Schreier, H., Repp, R., Rascher, W., et al. (2000) Nuclear gene targeting using negatively charged liposomes. *Int J Pharm*, **196**, 251–2.
24. Caracciolo, G., Pozzi, D., Capriotti, A.L., Marianecci, C., Carafa, M., et al. (2011) Factors determining the superior performance of lipid/DNA/protamine nanoparticles over lipoplexes. *J Med Chem*, **54**, 4160–71.
25. Chen, J., Yu, Z., Chen, H., Gao, J. and Liang, W. (2011) Transfection efficiency and intracellular fate of

Published by Woodhead Publishing Limited, 2013

polycation liposomes combined with protamine. *Biomaterials*, **32**, 1412–18.

26. Sun, X., Zhang, H.W. and Zhang, Z.R. (2009) Growth inhibition of the pulmonary metastatic tumors by systemic delivery of the p27 kip1 gene using lyophilized lipid-polycation-DNA complexes. *J Gene Med*, **11**, 535–44.
27. Liu, J., Guo, S., Li, Z., Liu, L. and Gu, J. (2009) Synthesis and characterization of stearyl protamine and investigation of their complexes with DNA for gene delivery. *Colloids Surf B Biointerfaces*, **73**, 36–41.
28. Yuan, H., Zhang, W., Du, Y.Z. and Hu, F.Q. (2010) Ternary nanoparticles of anionic lipid nanoparticles/protamine/DNA for gene delivery. *Int J Pharm*, **392**, 224–31.
29. Ma, Z., Zhang, J., Alber, S., Dileo, J., Negishi, Y., et al. (2002) Lipid-mediated delivery of oligonucleotide to pulmonary endothelium. *Am J Respir Cell Mol Biol*, **27**, 151–9.
30. Harvie, P., Dutzar, B., Galbraith, T., Cudmore, S., O'Mahony, D., et al. (2003) Targeting of lipid-protamine-DNA (LPD) lipopolyplexes using RGD motifs. *J Liposome Res*, **13**, 231–47.
31. Arangoa, M.A., Duzgunes, N. and Tros de Ilarduya, C. (2003) Increased receptor-mediated gene delivery to the liver by protamine-enhanced-asialofetuin-lipoplexes. *Gene Ther*, **10**, 5–14.
32. Maruyama, K., Iwasaki, F., Takizawa, T., Yanagie, H., Niidome, T., et al. (2004) Novel receptor-mediated gene delivery system comprising plasmid/protamine/sugar-containing polyanion ternary complex. *Biomaterials*, **25**, 3267–73.
33. Sun, X., Hai, L., Wu, Y., Hu, H.Y. and Zhang, Z.R. (2005) Targeted gene delivery to hepatoma cells using

Published by Woodhead Publishing Limited, 2013

galactosylated liposome-polycation-DNA complexes (LPD). *J Drug Target*, **13**, 121–8.
34. Mathews, M.S., Shih, E.C., Zamora, G., Sun, C.H., Cho, S.K., et al. (2012) Glioma cell growth inhibition following photochemical internalization enhanced non-viral PTEN gene transfection. *Lasers Surg Med*, **44**, 746–54.
35. Ye, J., Wang, A., Liu, C., Chen, Z. and Zhang, N. (2008) Anionic solid lipid nanoparticles supported on protamine/DNA complexes. *Nanotechnology*, **19**, 285708.
36. Delgado, D., Gascon, A.R., Del Pozo-Rodriguez, A., Echevarria, E., Ruiz de Garibay, A.P., et al. (2012) Dextran-protamine-solid lipid nanoparticles as a non-viral vector for gene therapy: in vitro characterization and in vivo transfection after intravenous administration to mice. *Int J Pharm*, **425**, 35–43.
37. Delgado, D., del Pozo-Rodriguez, A., Solinis, M.A. and Rodriguez-Gascon, A. (2011) Understanding the mechanism of protamine in solid lipid nanoparticle-based lipofection: the importance of the entry pathway. *Eur J Pharm Biopharm*, **79**, 495–502.
38. Du, Y.Z., Lu, P., Yuan, H., Zhou, J.P. and Hu, F.Q. (2011) Quaternary complexes composed of plasmid DNA/protamine/fish sperm DNA/stearic acid grafted chitosan oligosaccharide micelles for gene delivery. *Int J Biol Macromol*, **48**, 153–9.
39. Vighi, E., Montanari, M., Ruozi, B., Iannuccelli, V. and Leo, E. (2012) The role of protamine amount in the transfection performance of cationic SLN designed as a gene nanocarrier. *Drug Deliv*, **19**, 1–10.

Published by Woodhead Publishing Limited, 2013

13

Dendrimers

DOI: 10.1533/9781908818645.259

Abstract: Dendrimers are well-defined, multivalent molecules having branched structure of nanometer size. Dendrimers possess a distinct molecular architecture that consists of three different domains: (i) a central core (ii) branches (iii) terminal functional groups, present at the outer surface of the macromolecule, that dictates the nucleic acid complexation or drug entrapment efficacy. Dendrimers of defined size and structure can be engineered by a step-wise chemical synthesis approach, and that too with low polydispersity index. Generally, dendrimers are synthesized by two conceptually different approaches, the divergent and the convergent. Over the past few years, dendrimers have been attractive candidates for *in vitro* and *in vivo* gene delivery. The present chapter explores the application of dendrimers for delivering various genes to *in vitro* and *in vivo* milieux.

Key words: dendrimers, convergent, divergent, protonation, PAMAM dendrimers, PPI dendrimers, dendritic, triazine.

13.1 Introduction

Dendrimers (derived from the Greek "dendron," which means tree, and "meros," meaning part) are well-defined,

Published by Woodhead Publishing Limited, 2013

multivalent molecules having a branched structure (Figure 13.1). These molecules were first reported by Vögtle et al. as "cascade molecules," and later Tomalia et al. named these molecules dendrimers.[1,2] Dendrimers are globular macromolecules of nanometer size and possess a distinct molecular architecture that consists of three different domains: (i) a central core that comprises either a single atom or a group of atoms having at least two chemical functionalities that provide linkage for the branches; (ii) branches that emerge from the core, comprising repeating units with at least one junction of branching, whose repetition is organized in a geometric progression that results in a series of radially concentric layers termed generations (G); and (iii) terminal functional groups, present at the outer surface of the macromolecule, that dictate the efficacy of nucleic acid complexation or drug entrapment. Interactions of dendrimers with solvents, surfaces, or other molecules are due to the presence of a large number of terminal groups. Usually, the presence of highly interactive functionalities on the surface of dendrimers confers on them high reactivity, solubility, and binding properties.

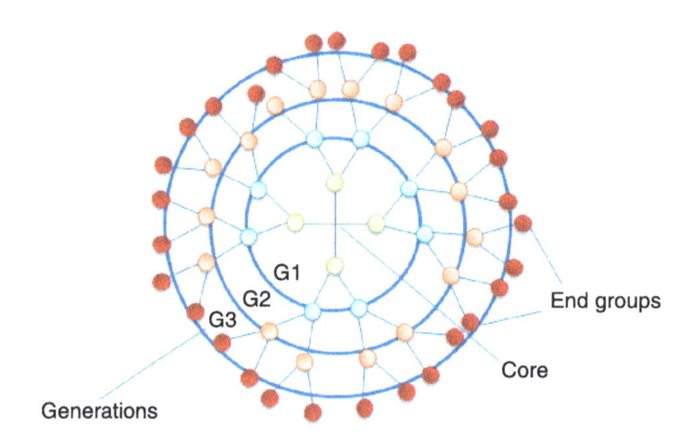

Figure 13.1 Structure of dendrimers

Published by Woodhead Publishing Limited, 2013

Dendrimers of defined size and structure and with a low polydispersity index (PDI) can be engineered by a step-wise chemical synthesis approach. The synthesis approaches are quite versatile in the sense that a wide range of dendrimer molecules can be generated, with different functionalities. Generally, dendrimers are synthesized by two conceptually different approaches, the divergent and the convergent (Figure 13.2). The divergent approach consists of dendrimer synthesis starting from the multifunctional core, followed by building one monomer layer or generation at a time.[3] The core molecule reacts with monomer molecules containing one reactive group and two (or more) inactive groups. In contrast, the convergent approach consists of dendrimer synthesis starting from the end groups and terminating at the core.[2] In this approach two (or more) outer surface branch

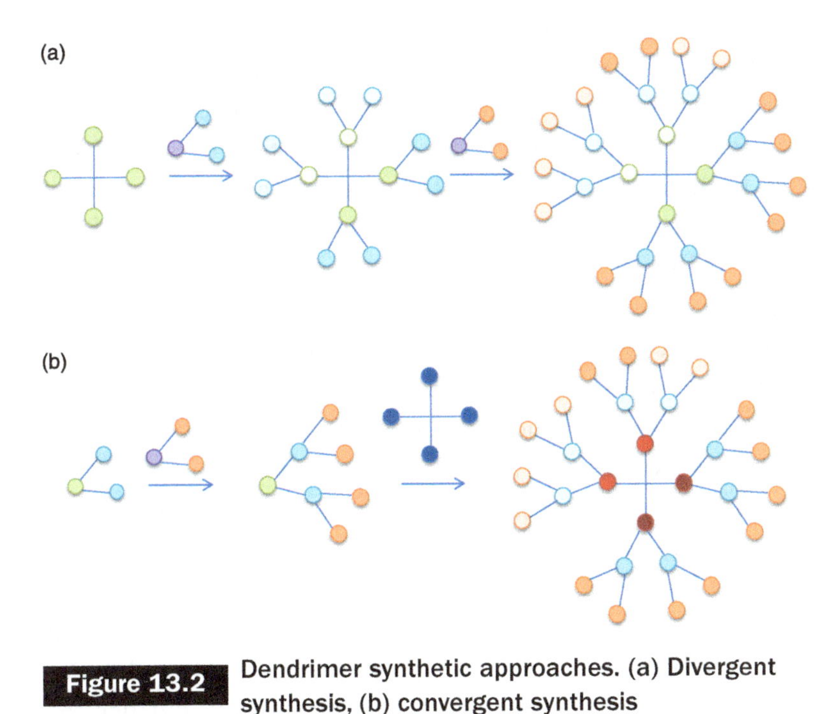

Figure 13.2 Dendrimer synthetic approaches. (a) Divergent synthesis, (b) convergent synthesis

Published by Woodhead Publishing Limited, 2013

subunits are reacted with a single joining unit which has two (or more) corresponding active sites and a distal inactive site.

13.2 Dendrimers in gene delivery

Over the past few years, dendrimers have been attractive candidates for gene delivery. Dendrimers form complexes with DNA through sequence-independent electrostatic interactions between negatively charged phosphate groups of the nucleic acid and protonated (positively charged) primary amino groups on the dendrimer surface, resulting in particles termed dendriplexes. Some of the advantages associated with dendrimers are: i) precise control over size and structure, ii) high chemical and structural homogeneity, and iii) high ligand and functional groups distribution. All of these can be manipulated as per the delivery application. The biocompatibility of dendrimers is influenced by size, surface charge, and concentration; dendrimers do not exert significant cytotoxicity *in vitro* or *in vivo*.[4–11] Further, PAMAM dendrimers have also been suggested to mediate transport through the epithelial barrier, indicating their potential for oral delivery.[12–14] The morphology of dendrimers is significantly influenced by both the core and the number/type of interior branching units. The steric crowding occurs at the surface of dendrimers, since the diameter of dendrimers increases linearly whereas the number of surface groups increases exponentially for each generation. Therefore, the dendrimers at higher generations form more compact, rigid three-dimensional structures while low generations yield flexible and open structures.[15] For instance, PAMAM dendrimers become more rigid and compact at generation 4.5, and in the case of polypropylenimine (PPI) globular structures are observed at the fourth generation. Further, the

Published by Woodhead Publishing Limited, 2013

globular nature of the higher-generation dendrimers results in different insolubility and reactivity of the end groups in comparison to similar linear polymer analogues. Due to these physico-chemical properties, usually fourth-generation or higher dendrimers are employed in therapeutic applications.

Dendrimer conformations are influenced by ionic strength and pH, when suspended in aqueous solutions, with the intensity of changes depending on the type of functional group at the dendrimer surface. Molecular dynamic studies of amine-terminated PAMAM dendrimers revealed that at high pH globular loosely compact structures are observed, whereas the extended conformation dominates at low pH (~5) due to electrostatic repulsion between the protonated tertiary amines (pKa ~5) at the interior of the dendrimer and the primary amines (pKa ~9–11) at the surface.[16] This alteration in conformation influences the endosomal escape potential of dendrimers, followed by cellular internalization. At physiological pH (7.4) only the primary amines are protonated, while exposure to low pH (~5) in the endosomal compartment leads to protonation of the tertiary amines, thereby resulting in endosomal rupture and subsequent release of dendrimers into the cytoplasm. On the other hand, the carboxyl-terminated PPI dendrimers display an extended conformation at both low (~4) and high (~11) pH due to protonation of amines at low pH or deprotonation of carboxylic acids at high pH. Further, the PPI dendrimer showed a condensed, back-folded structure at pH ~6 due to intramolecular hydrogen bonding of the zwitterionic structure.[17]

13.2.1 Polyamidoamine dendrimers

One of the initial studies on the use of Starburst PAMAM dendrimers observed high transfection efficiency in a variety

Published by Woodhead Publishing Limited, 2013

of suspension and adherent cultures of mammalian cells, using luciferase and β-galactosidase as reporter plasmids.[18] The dendrimers of generation six (G6) with NH_3 core showed the best efficiency at a N/P ratio of 6 and dendrimers of size ~6.8 nm. When GALA, a water-soluble, membrane-destabilizing peptide, was covalently attached to the dendrimer via a disulfide linkage, transfection efficiency of the 1:1 complex increased by two to three orders of magnitude. It was argued that PAMAM dendrimers had the ability to escape from the endosome due to the ability of the internal dendrimer amine groups to buffer pH changes in the endosome by a mechanism similar to that described for PEIs.[18] Later, PAMAM dendrimers were observed to efficiently transfect a wide variety of cells with minimal cytotoxicity.[5] However, for efficient transfection additional agents were sometimes required, such as diethylaminoethyl (DEAE) dextran, which appeared to alter the nature of the complex, or chloroquine, which acted as a lysosomotropic agent. The optimal generation of PAMAM dendrimers ranged from G7 to G10, which showed transfection in most of the cell lines.[5] Further, the cytotoxicity of PAMAM dendrimers (G3, G5 and G7) was investigated both *in vitro* and *in vivo*.[7] *In vitro* cell viability of V79 cells exposed to dendrimers was found to be generation-dependent; PAMAM G3 affected cell growth at 1 mM whereas PAMAM G5 showed cytotoxicity at 10 μM. In both cases, cell viability was less than 10%. Cell growth was greatly affected by PAMAM G7, with which less than 10% cell viability was obtained even at a low concentration of 100 nM. Further, *in vivo* studies did not reveal any significant cytotoxicity or immune response. Biodistribution studies of PAMAM dendrimers showed the highest accumulation in kidney for G3, while G5 and G7 localized in the pancreas. Additionally, G7 dendrimers showed extremely high urinary excretion.[7]

PAMAM dendrimers were also employed for delivery of AS-ODNs and "antisense expression plasmids" for the targeted modulation of gene expression.[6] Dendrimer-mediated delivery of AS-ODNs or antisense cDNA plasmids into various cell lines resulted in a specific and dose-dependent inhibition of luciferase expression. A 25–50% reduction of baseline luciferase activity, along with improvement in cell viability, was observed for ODNs complexed to dendrimers as compared with uncomplexed ODNs.[6] Further, cellular uptake of both dendrimers and ODNs was investigated after incubation of Oregon green 488-conjugated dendrimer-carboxytetramethylrhodamine (TAMRA)-oligonucleotide complexes with HeLa cells in culture.[19] A remarkably significant number of cells showed strong fluorescence intensity in the nuclei. Furthermore, a dramatic increase in the luciferase expression level was observed when AS-ODNs were delivered with the Oregon green-dendrimer conjugates, as compared with unmodified PAMAM dendrimers. Cell viability was also enhanced when dendriplexes were treated in the presence of serum, while cell viability was reduced for dendrimer alone or the dendriplexes in serum-free medium.[19] The influence of anionic oligomers (ODNs, dextran sulfate) on the transfection efficacy of PAMAM dendriplexes was evaluated in HeLa and NIH3T3 cells.[20] The transfection efficacy of dendriplexes was significantly enhanced when anionic oligomers were mixed with pDNA before addition of the dendrimer. This improvement in transfection efficiency was dependent on the size, structure, and charge of oligomers, with 35–50 ODNs being most efficient.[20]

To improve gene delivery efficiency, quaternary ammonium salts have been introduced to the internal tertiary amine of surface hydroxyl-terminated PAMAM dendrimer (PAMAM-OH) and acetylated PAMAM dendrimer (PAMAM-NHAc). The morphology and size of the

Published by Woodhead Publishing Limited, 2013

complexes was dependent on the degree of quaternization; 0.52 QPAMAM-OH (52% degree of quaternization) showed partially condensed structures of DNA, while small and spherical particles were observed with 0.78 QPAMAM-OH and 0.97 QPAMAM-OH (78 and 97% degree of quaternization, respectively).[21] The transfection efficiency in HEK293T cells of QPAMAM-OH was lower than that of PAMAM G4 and PEI.[21] In another study, the levels of luciferase expression mediated by 0.5 QPAMAM and 0.7 QPAMAM were lower than those of 0.1 QPAMAM (10% degree of quaternization) and unmodified dendrimers.[22] The expression levels mediated by 0.5-QPAMAM and 0.7-QPAMAM exhibited a tenfold decrease as compared with that of unmodified dendrimer.[22] The decrease in the surface charge of PAMAM dendrimers resulted in reduced interactions with the cell membrane, thus causing lower transfection efficiencies. However, quaternization improved the cell viability as compared with the unmodified PAMAM dendrimers. A positive correlation between cell viability and degree of quaternization was detected; PAMAM < 0.1-QPAMAM < 0.5-QPAMAM < 0.7-QPAMAM, with cell viability according to the degree of quaternization.[22] A similar trend was observed for quaternized PAMAM-OH dendrimers, for which the relative cell viability for quaternized dendrimers was over 90% at all concentrations tested (up to 100 μg/ml).[21] The higher cell viability may be attributed to the shielding of the interior positive charges by surface hydroxyl groups.

PAMAM dendrimers can be activated by hydrolytic cleavage of some of the amide bonds in the inner part of the polymer and removal of some of the dendrimer branches.[23] To perform hydrolytic cleavage of the amide bonds, dendrimers are solubilized in an appropriate solvolytic solvent and heated for a defined time period. Dendrimer

Published by Woodhead Publishing Limited, 2013

degradation is a random process, leading to a population of mixed compounds with MW ranging from very low (<1500 Da) to several kilodaltons. Carboxyl groups are formed at the cleavage sites, and dendrimers thus formed possess a higher degree of flexibility. However, the overall size and shape of the dendrimer remain constant following the activation process.[23] The differences between dendriplexes prepared with activated and non-activated PAMAM dendrimers G4 and G5 revealed that particle sizes for non-activated dendriplexes remained the same, whereas for activated dendriplexes the sizes decreased slightly in both G4 and G5 dendriplexes with increasing N/P.[24] However, the zeta potentials generally increased with increasing N/P ratio, especially for G5 dendriplexes, for both non-activated and activated dendriplexes. No significant difference in the surface charge of dendriplexes was observed by adjusting the activation time or by using a different dendrimer generation.[24]

Activated PAMAM dendrimers have been observed to mediate higher transfection efficiency as compared with non-activated ones. A study involving PAMAM G6 and G8 dendrimers based on ethylene diamine core (6EDA and 8EDA, respectively) as well as PAMAM G5 and G6 dendrimers based on tris(2-aminoethyl)amine core (5TAEA and 6TAEA, respectively) suggested the existence of an optimal heat treatment time for achieving maximum transfection efficiency.[23] The lower-generation dendrimers required a short heat treatment duration to achieve maximum transfection efficiency, although their corresponding efficiencies were lower than those of higher-generation dendrimers. The transfection efficiencies of activated 6TAEA dendriplex was enhanced by about 100- and 1000-fold as compared with that of non-activated dendriplex for β-galactosidase and luciferase assays, respectively.[23] Another study investigated the influence of activation time on

Published by Woodhead Publishing Limited, 2013

dendrimer activation and associated flexibility for transfection.[24] Maximum transfection efficiency observed for activated dendriplex based on PAMAM G4 was about fourfold (N/P = 4 and 6) higher than that for the corresponding non-activated dendriplex. For activated dendriplex based on PAMAM G5, the efficiency was about twofold (N/P = 4) and sevenfold (N/P = 6) higher than for the corresponding non-activated dendriplex.[24] *In vivo* studies on mice also observed the same trend, especially for the transfection efficiency obtained in the lungs. These results suggested that both generation number and duration of heat treatment can be optimized to enhance transfection efficacy of PAMAM dendrimers. Changes in global gene-expression profiles in HeLa cells exposed to non-activated and activated PAMAM dendrimers, alone or in complexes with pDNA (dendriplexes), were investigated.[25] Real-time quantitative RT-PCR confirmed four regulated genes (PHF5A, ARNTL2, CHD4, and P2RX7) affected by activated dendrimers and dendriplexes. Activated and non-activated dendrimers and dendriplexes alike induced multiple gene expression changes, some of which overlapped with their dendriplexes. Dendrimers and dendriplexes principally affect genes with the molecular functions of nucleic acid binding and transcription activity, metal-ion binding, enzyme activity, receptor activity, and protein binding.[25]

The transfection efficiency of PAMAM dendrimers has been improved by attachment of different ligands. Coupling of dendrimers with cell-penetrating peptides such as arginine has been observed to enhance the transfection efficiency. An arginine-grafted-PAMAM dendrimer (PAMAM-Arg) was designed that consists of about 58 out of 64 surface primary amine groups coupled with arginine residues on PAMAM G4.[26] The transfection efficiency of PAMAM-Arg-DNA complex was more than 10- and 100-fold higher than that of

Published by Woodhead Publishing Limited, 2013

unmodified dendrimer-DNA complex in HepG2 and Neuro 2A cells respectively. Further, studies on HeLa, HEK 293, and A549 cells showed that complexes formed by di-arginine-conjugated dendrimers generally led to higher transfection efficiencies than mono-arginine-conjugated dendrimers.[27] Intracellular trafficking showed that all the complexes were internalized into the cytosol, but only complexes formed by di-arginine-conjugated PAMAM G3 and G4 were localized within the nuclei. However, the cell viability of di-arginine-conjugated dendrimers was poor, comparable to PEI.[27]

To improve transfection efficiency, a series of PAMAM G2 dendrimers were conjugated with α-, β- and γ-cyclodextrins (CD conjugates) and evaluated in NIH3T3 and RAW264.7 cells.[28] The α-CD conjugate showed higher transfection efficiency than β- and γ-CD conjugates. Also, the gene expression of α-CD conjugate-DNA complexes was significantly higher than that of the dendriplexes over the studied range of N/P ratios. At high N/P ratio, α-CD conjugate mediated higher transfection efficiency than Lipofectin, but lower than that of TransFast.[28] Further, the influence of generation of dendrimers (G2, G3, and G4) on transfection efficiencies was investigated for CD conjugates.[29] For both NIH 3T3 and RAW264.7 cells, α-CD conjugate of PAMAM G3 mediated the highest transfection efficiency as compared with α-CD conjugate of PAMAM G2 and G4, and was quite comparable to TransFast. Further, the effect of degree of substitution of α-CD in CD conjugates based on PAMAM G3 dendrimer was investigated with 1.1, 2.4 and 5.4 degrees of substitution.[29] The α-CD conjugates with high degrees of substitution (2.4 and 5.4) showed membrane-disruptive properties, and 5.4 substitution resulted in significant cytotoxicity at high N/P ratios.[30] The α-CD conjugate with degree of substitution of 2.4 exhibited

Published by Woodhead Publishing Limited, 2013

superior transfection efficiency in NIH3T3 and HepG2 cells at high N/P ratio in comparison to other conjugates. *In vivo* studies on BALB/c mice followed the trend; α-CD conjugate with degree of substitution of 2.4 showed higher transfection efficiency as compared with unmodified dendrimer, and α-CD conjugates localized in spleen at 12 h and in liver and kidney at 24 h post-administration.[30] These studies suggested that CD conjugation to dendrimers influences the intracellular trafficking of DNA, possibly in terms of increased release of DNA from the endosome into the cytoplasm after endocytosis.

The delivery of PAMAM dendriplexes into the cell nucleus has been enhanced via glucocorticoid conjugation. The transfection efficacy of PAMAM-dexamethasone (PAMAM-Dexa) (four dexamethasone residues on PAMAM G4) was evaluated in HEK 293 and Neuro2A cells.[31] In the absence of serum, the transfection efficiency of PAMAM-Dexa-DNA complex was found to be 2-fold and 6-fold higher than that of unmodified dendriplexes in HEK293 and Neuro 2A cells, respectively. However, in the presence of serum, the transfection efficiency of PAMAM-Dexa-DNA complex was 10-fold better than that of PEI/DNA complexes in HEK293 and Neuro 2A cells, respectively. Further, confocal microscopy images confirmed localization of more PAMAM-Dexa-DNA complex in the nucleus region than for unmodified dendrimer-DNA complex.[31] In a similar study, a glucocorticoid of high potency, triamcinolone acetonide, was conjugated to PAMAM dendrimer to form PAMAM-TA with different degrees of substitution, mainly PAMAM-TA-L (0.22 triamcinolone acetonide residues on PAMAM G4) and PAMAM-TA-H (1.65 triamcinolone acetonide residues on PAMAM G4).[32] Enhanced transfection efficiency was observed in both HEK 293 and HepG2 cells, irrespective of presence of serum, as compared with unmodified dendriplexes, though there was no significant difference in

Published by Woodhead Publishing Limited, 2013

expression between PAMAM-TA-L and PAMAM-TA-H at the respective optimal weight ratio. Intracellular localization experiments suggested efficient translocation of PAMAM-TA from the cytoplasm into the cell nucleus.[32]

13.2.2 Polypropylenimine dendrimers

Polypropylenimine dendrimers contain 100% protonatable amine groups, which renders them suitable for DNA binding and delivery.[33] PPI dendrimers derived from a 1,4-diaminobutane core with different generations (DAB 4, 8, 16, 32, and 64) have been employed for gene delivery.[34] DNA binding and cytotoxicity were generation-dependent and followed the trend DAB 64 > DAB 32 > DAB 16 > DOTAP > DAB 4 > DAB 8. A different pattern was observed for transfection studies in human epidermoid carcinoma cell line (A431 cells): DAB 8 = DOTAP = DAB 16 > DAB 4 > DAB 32 = DAB 64.[34] Further, quaternization of the lower-generation DAB 4 and 8 dendrimers significantly improved DNA binding and biocompatibility.[35] DNA binding decreased the toxicity of the low-generation dendrimers, whereas the toxicity of the higher generations (DAB 16 and 32) was not altered. The DAB 8/pDNA formulations were lethally toxic, whereas quaternized DAB 8/pDNA complexes were tolerated on intravenous injection. Quaternization improved cell biocompatibility of higher DAB/DNA dendrimer complexes, but reduced transfection efficiency of higher-generation native dendrimers. DAB 16 and quaternized DAB 8/DNA formulations led to liver-targeted gene expression, presumably in Kupffer cells, and avoided lung delivery, in contrast to LPEI.[35]

Toxicogenomic examination of generation 2 (DAB-8) and generation 3 (DAB-16) PPI dendrimers in two human cell lines revealed that PPI dendrimers alone elicited marked

Published by Woodhead Publishing Limited, 2013

changes in endogenous gene expression in A431 epithelial cells.[36] Further, the extent of PPI-induced gene changes was dependent on the dendrimer generation, as the number of genes affected was greater with G3 than with G2 PPI dendrimers. Also, the signature of DAB16-induced gene expression changes in A549 cells was different from those elicited in A431 cells, suggesting a strong dependence on cell type. In addition to this, modulation in expression of a variety of gene ontologies, including those involved in defense responses, cell proliferation, and apoptosis, was observed.[36] Hence, it was deduced that dendrimers can intrinsically manipulate the expression of many endogenous genes, which could potentially lead to their exerting multiple biological effects in cells.[36]

Arginine-conjugated PPI G2 (DAB-8) dendrimers (PPI2-R) were engineered for gene delivery, complexed with DNA to form polyplexes of size ~200 nm at a N/P ratio of 150.[37] PPI2-R displayed 80–90% cell viability even at a concentration of 150 μg/ml. Moreover, the transfection efficiency of PPI2-R was found to be 8–214-fold higher than that of unmodified PPI2 in HeLa and 293 cells, and quite comparable to that of PEI (25 kDa). Furthermore, PPI2-R showed two to 3-fold higher transfection efficiency than PEI in human umbilical vein endothelial cell (HUVEC) cells, proving its potency as a gene delivery carrier for primary cells.[37] Low-generation PPI dendrimer (DAB-8) was conjugated to β-CD, and *in vitro* studies suggested low cytotoxicity and high transfection efficiency.[38]

PPI G2 or G3 was coupled with branched oligoethylenimine 800 Da (OEI) or PPI dendrimer via hexanediol diacrylate, providing bio-reversible ester linkages.[39] Grafted dendrimers efficiently condensed DNA to nanosized dendriplexes (100–200 nm) and exhibited an increased colloidal stability as compared with their unmodified counterparts. Transfection

Published by Woodhead Publishing Limited, 2013

efficacy of OEI-grafted dendrimer/pDNA dendriplexes was maximal, which was comparable to PEI/DNA polyplexes, suggesting that the incorporation of ethylenimine moieties is the key factor contributing to this enhanced transfection efficiency. Further, intravenous injection of OEI-grafted PPI polyplexes into tumor-bearing mice mediated transgene expression predominantly in subcutaneous tumors. Also, gene expression levels in tumors increased significantly with higher dendrimer core generation.[39]

PPI G3 modified with Pluronic P123 showed much lower cytotoxicity than that of PPI alone.[40] Physico-chemical studies showed that both P123-PPI and PPI condense pDNA into nanoparticles with a size of ~100 nm and a zeta potential of about +15 mV at the N/P ratio of 20. Further, the transfection efficiency of P123-PPI/DNA nanoparticles in SPC-A1 cells was much higher than that of unmodified PPI/DNA nanoparticles. The addition of free P123 during the preparation of P123-PPI/DNA nanoparticles significantly enhanced the transfection efficiency in the presence of 10% FBS.[40]

Studies of conjugation of transferrin to PPI dendrimer G3 (DAB-TF) revealed a slight increase of the polyplex size compared with the unmodified DAB polyplex, from ~196 nm to ~287 nm, with lower PDI.[41] Maximal gene expression levels of β-galactosidase were obtained at a N/P ratio of 10 in T98G and A431 cells, and were markedly higher than for the unmodified dendrimer and DOTAP. Moreover, grafting of TF to DAB improved DNA uptake by T98G cells, but had little influence on A431 cell line, for which the nuclear uptake of DNA appeared to be similar after treatment with both targeted and non-targeted dendrimers.[41] The intravenous administration of DAB-TF polyplexes resulted in gene expression mainly in the tumors. Subsequently, the intravenous administration of the polyplexes led to a rapid

Published by Woodhead Publishing Limited, 2013

and sustained tumor regression over 1 month, with long-term survival of 100% of the animals (90% complete response, 10% partial response). Additionally, the treatment was well tolerated by the animals, with no apparent signs of toxicity.[41] In another study, DAB-TF complexed to a plasmid DNA encoding p73 led to an enhanced anti-proliferative activity *in vitro*, by up to 120-fold in A431 compared with the unmodified dendriplex.[42] *In vivo* intravenous administration of this p73-encoding dendriplex resulted in a rapid and sustained inhibition of tumor growth over 1 month, with complete tumor suppression for 10% of A431 and B16-F10 tumors and long-term survival of the animals.[42]

PPI G4 was employed to target 45-base (phosphorothioate-modified) ODNs in the cell nuclei of cultured cells in order to correct a specific point mutation in the hypoxanthine-guanine-phosphoribosyl-transferase (*hprt*) gene in hamster fibroblasts (V79-400 cells).[43] Transfection resulted in a high cellular uptake of the FITC-labeled ODNs in V79-400 and HuH-7 cells, with low toxicity levels. Localization studies showed presence of ODNs 24 h after transfection predominantly in the nuclear compartment. Moreover, the ODNs mediated the desired biological activity, the correction of the *hprt* point mutation in V79-400 cells.[43]

13.2.3 Dendritic poly(L-lysine)

Dendritic poly(L-lysine) (DPK) was first synthesized by Dankewalter et al. using Boc-L-Lys(Boc)-OH benzhydrylamide as a core.[44] DPK can be synthesized as a well-defined molecule possessing a precise number of terminal amines per dendrimer. The initial studies reported use of several types of DPKs, block copolymers consisting of PEG and PLL dendrimer.[45,46] The copolymers formed spherical particles upon complexation with DNA of size ~100 nm and did not show any cytotoxity

Published by Woodhead Publishing Limited, 2013

toward NIH3T3 cells even at higher concentration.[46] Amphipathic asymmetric PLL dendrimers, in which the dendritic structure was attached to an amino group of α-amino myristic acid, have been used to transfect COS7 cells at a N/P ratio of 5.[47] These dendrimers further transfected BHK cells at a N/P ratio of 5, resulting in 20% efficiency. Also, cytotoxicity was observed to increase when the N/P ratio exceeded 5.[48]

Several generations of symmetric DPK prepared from hexamethylenediamine as an initiator core were investigated for their abilities binding pDNA and gene transfection into cells.[49] Dendrimers of generations 3 to 6 yielded particles of size ~200 to 250 nm. Efficient gene transfection without loss of cell viability was observed into several different cell lines with DPKs of generations 5 and 6, which have 64 and 128 amine groups on the surface of the molecule, respectively. Additionally, the transfection efficiency of the DPK of the sixth generation remained unaltered even in the presence of 50% serum in the transfection medium.[49]

The zeta potential of KG6/DNA complexes was maintained at a neutral or slightly positive value (+3 mV) even when the N/P ratio was increased to 8, which enhanced stability in the presence of a high concentration of serum. In addition, the complexes were found to circulate in the bloodstream for at least 3 h after intravenous injection into mice.[50] This prolonged blood circulation was due to the stealth character provided by neutral and hydrophilic surfaces.

The effect of substituting terminal cationic groups on the KG6 dendrimer was investigated by replacing terminal amino acids by arginine (KGR6) and histidine (KGH6).[51] The KGR6 dendrimer could bind to the pDNA as strongly as KG6, whereas KGH6 showed decreased binding ability. Further, KGR6 showed 3–12-fold higher transfection efficiency than KG6 into several cell types, while transfection

Published by Woodhead Publishing Limited, 2013

was absent with KGH6. However, complexation of KGH6 with DNA under acidic conditions (pH 5.0) followed by transfection showed better efficiency than that with KG6.[51]

A comparison between DPK generation 6 (KG6) and PLL showed that KG6 formed a neutral DNA complex, and its DNA compaction level was weaker than that of PLL.[52] Further, uptake of complexes into cells mediated by PLL was fourfold higher than that with KG6. However, gene expression mediated by KG6 was 100-fold higher than that by PLL. This improved transfection efficacy was due to higher proton sponge effect and weak compaction of DNA by KG6 that led to escape from endosomal degradation and allowed access to RNA polymerase in the cell nucleus. Intravenous administration of KG6 and PLL in mice did not show a significant difference in cytokine production.[52]

Amphiphilic DPKs of generation 4 (KG4) condensed DNA to form particles of size ~100–200 nm, and the zeta potential gradually changed from negative to positive values as the N/P ratio increased from 1 to 25.[53] The higher-generation dendrimers produced higher positive zeta potentials, suggesting a stronger potency of the complexes to interact with negatively charged cell membranes. Good biocompatibility of the dendrimers and their complexes was observed by conducting *in vitro* and *in vivo* cytotoxicity assays over the different N/P ratios studied. Further, the transfection efficiency of KG5 was higher than other dendrimers and insensitive to variation in serum concentration.[53]

13.2.4 Triazene dendrimers

Triazine dendrimers are synthesized by sequential substitution of trichlorotriazine with amine nucleophiles. The variance in the reactivity of different chlorotriazines (mono-, di-, and trich-loro) facilitates high-yielding, selective synthesis across a range of nucleophiles through modulation of temperature and reaction times.[54,55] The gene delivery efficacy of triazine dendrimers depends on the size, peripheral group, and internal structure of dendrimers, which significantly influence the condensation of ODNs and their delivery to target cells. The nomenclature of these dendrimers indicates structure, size, and surface chemistry (e.g. *RG-G1-A: RG* for rigid, *BT* for bow-tie, *FX* for flexible, *Gn* for generation, and *A–K* for surface groups). To evaluate the effect of the peripheral groups on transfection efficiency, a small library of eight rigid triazine dendrimers (all second-generation) with amine (*A*), hydroxyl (*H*), guanidine (*G*), and aliphatic chain surface groups (*F*) was prepared.[56] The DNA condensation was dependent on the peripheral groups of the dendrimers. Dendrimers with only six primary amines on the surface (*RG-G2-H* and *RG-G2-I*) showed less DNA compaction as compared with dendrimers with 12 amine groups (*RG-G2-A* and *RG-G2-F*), resulting in 60–70% and less than 30% of free or loosely bound DNA, respectively, at a N/P ratio of 5. A similar trend was observed in the guanidinylated dendrimers (*RG-G2-B*, *RG-G2-J*, *RG-G2-K*), and they appeared most effective in packing DNA in contrast to the dendrimers with the same number of protonatable amines.[56] Small complexes (≤ 200 nm) were formed by *RG-G2-A* and *RG-G2-E* upon complexation with DNA, probably due to their high positive surface charge, which may facilitate cellular internalization via endocytosis. Transfection experiments in human melanoma cell lines (MeWo) suggested that *RG-G2-A* dendriplex possessed maximal transfection efficiency at N/P ratios of 5 and 7. Further, no significant cytotoxicity was associated with the dendrimers (IC_{50} 0.03–0.06 mg/ml).[56]

Another study employed the surface group *A*, with generations (*G1–G3*) and structure: *RG-G1-A*, *RG-G2-A*, *RG-G3-A*, *BT-G2-A*, and *FX-G2-A*.[57] The degree of DNA

Published by Woodhead Publishing Limited, 2013

compaction was generation-dependent, increasing with higher generations of the rigid dendrimers, while dendrimer structure had no effect. For instance, *BT-G2-A* and *FX-G2-A* formed looser dendriplexes than *RG-G2-A*, presumably due to the lower charge density (~262 Da/N atom). Although the cytotoxicity of the rigid dendrimers increased with increasing generation (IC_{50} 20–500 μg/ml), it was significantly lower than for PEI 25 kDa (IC_{50} 4.9 μg/ml).[57] Further, the transfection efficiency of the rigid dendriplexes increased with increasing generation in both L929 and MeWo cells. The *FX-G2-A* dendriplex showed maximal transfection efficiency as compared with other dendriplexes studied, and controls including PEI, PAMAM, and SuperFect.[57]

13.3 Conclusions

One of the major advantages associated with dendrimers is their synthetic versatility, derived from sequential reactivity. Recent progress in dendrimer syntheses, including availability of higher generations (up to 11) and adoption of new linkers, makes the future of the dendrimer platform as a gene delivery vector promising. However, synthetic challenges include increasingly rigorous methods to eliminate composition and batch-to-batch variability in larger-scale synthesis of higher-generation materials. PEGylation introduces significant heterogeneity into the dendrimers; hence, finding an alternative to PEG will also be one of the major challenges.

13.4 References

1. Buhleier, E., Wehner, W. and Vögtle, F. (1978) "Cascade"- and "Nonskid-Chain-like" Syntheses of Molecular Cavity Topologies. *Synthesis*, **1978**, 155,158.

Published by Woodhead Publishing Limited, 2013

2. Tomalia, D.A., Baker, H., Dewald, J., Hall, M., Kallos, G., et al. (1985) A new class of polymers: starburst-dendritic macromolecules. *Polym J*, **17**, 117–32.
3. Hawker, C.J. and Frechet, J.M.J. (1990) Preparation of polymers with controlled molecular architecture. A new convergent approach to dendritic macromolecules. *J Am Chem Soc*, **112**, 7638–47.
4. Qin, L., Pahud, D.R., Ding, Y., Bielinska, A.U., Kukowska-Latallo, J.F., et al. (1998) Efficient transfer of genes into murine cardiac grafts by Starburst polyamidoamine dendrimers. *Hum Gene Ther*, **9**, 553–60.
5. Kukowska-Latallo, J.F., Bielinska, A.U., Johnson, J., Spindler, R., Tomalia, D.A., et al. (1996) Efficient transfer of genetic material into mammalian cells using Starburst polyamidoamine dendrimers. *Proc Natl Acad Sci U S A*, **93**, 4897–902.
6. Bielinska, A., Kukowska-Latallo, J.F., Johnson, J., Tomalia, D.A. and Baker, J.R., Jr. (1996) Regulation of in vitro gene expression using antisense oligonucleotides or antisense expression plasmids transfected using starburst PAMAM dendrimers. *Nucleic Acids Res*, **24**, 2176–82.
7. Roberts, J.C., Bhalgat, M.K. and Zera, R.T. (1996) Preliminary biological evaluation of polyamidoamine (PAMAM) Starburst dendrimers. *J Biomed Mater Res*, **30**, 53–65.
8. Duncan, R. and Izzo, L. (2005) Dendrimer biocompatibility and toxicity. *Adv Drug Deliv Rev*, **57**, 2215–37.
9. El-Sayed, M., Ginski, M., Rhodes, C. and Ghandehari, H. (2002) Transepithelial transport of poly(amidoamine) dendrimers across Caco-2 cell monolayers. *J Control Release*, **81**, 355–65.

Published by Woodhead Publishing Limited, 2013

10. El-Sayed, M., Rhodes, C.A., Ginski, M. and Ghandehari, H. (2003) Transport mechanism(s) of poly (amidoamine) dendrimers across Caco-2 cell monolayers. *Int J Pharm*, **265**, 151–7.
11. Jevprasesphant, R., Penny, J., Jalal, R., Attwood, D., McKeown, N.B., et al. (2003) The influence of surface modification on the cytotoxicity of PAMAM dendrimers. *Int J Pharm*, **252**, 263–6.
12. Jevprasesphant, R., Penny, J., Attwood, D., McKeown, N.B. and D'Emanuele, A. (2003) Engineering of dendrimer surfaces to enhance transepithelial transport and reduce cytotoxicity. *Pharm Res*, **20**, 1543–50.
13. Kitchens, K.M., El-Sayed, M.E. and Ghandehari, H. (2005) Transepithelial and endothelial transport of poly (amidoamine) dendrimers. *Adv Drug Deliv Rev*, **57**, 2163–76.
14. Wiwattanapatapee, R., Carreno-Gomez, B., Malik, N. and Duncan, R. (2000) Anionic PAMAM dendrimers rapidly cross adult rat intestine in vitro: a potential oral delivery system? *Pharm Res*, **17**, 991–8.
15. Bosman, A.W., Janssen, H.M. and Meijer, E.W. (1999) About dendrimers: structure, physical properties, and applications. *Chem Rev*, **99**, 1665–88.
16. Lee, I., Athey, B.D., Wetzel, A.W., Meixner, W. and Baker, J.R. (2002) Structural molecular dynamics studies on polyamidoamine dendrimers for a therapeutic application: effects of pH and generation. *Macromolecules*, **35**, 4510–20.
17. Rietveld, I.B., Bouwman, W.G., Baars, M.W.P.L. and Heenan, R.K. (2001) Location of the outer shell and influence of pH on carboxylic acid-functionalized poly(propyleneimine) dendrimers. *Macromolecules*, **34**, 8380–3.

Published by Woodhead Publishing Limited, 2013

18. Haensler, J. and Szoka, F.C. (1993) Polyamidoamine cascade polymers mediate efficient transfection of cells in culture. *Bioconjug Chem*, **4**, 372–9.
19. Yoo, H. and Juliano, R.L. (2000) Enhanced delivery of antisense oligonucleotides with fluorophore-conjugated PAMAM dendrimers. *Nucleic Acids Res*, **28**, 4225–31.
20. Maksimenko, A.V., Mandrouguine, V., Gottikh, M.B., Bertrand, J.-R., Majoral, J.-P., et al. (2003) Optimisation of dendrimer-mediated gene transfer by anionic oligomers. *J Gene Med*, **5**, 61–71.
21. Lee, J.H., Lim, Y.-B., Choi, J.S., Lee, Y., Kim, T.-I., et al. (2003) Polyplexes assembled with internally quaternized PAMAM-OH dendrimer and plasmid DNA have a neutral surface and gene delivery potency. *Bioconjug Chem*, **14**, 1214–21.
22. Lee J.H., Lim Y.B., Choi J.S., Choi M.U., Yang C.H., et al. (2003) Quaternized polyamidoamine dendrimers as novel gene delivery system: relationship between degree of quaternization and their influences. *Bull Korean Chem Soc*, **24**, 1637–40.
23. Tang, M.X., Redemann, C.T. and Szoka, F.C., Jr. (1996) In vitro gene delivery by degraded polyamidoamine dendrimers. *Bioconjug Chem*, **7**, 703–14.
24. Navarro, G. and Tros de Ilarduya, C. (2009) Activated and non-activated PAMAM dendrimers for gene delivery in vitro and in vivo. *Nanomedicine*, **5**, 287–97.
25. Kuo, J.H., Liou, M.J. and Chiu, H.C. (2010) Evaluating the gene-expression profiles of HeLa cancer cells treated with activated and nonactivated poly(amidoamine) dendrimers, and their DNA complexes. *Mol Pharm*, **7**, 805–14.
26. Choi, J.S., Nam, K., Park, J.Y., Kim, J.B., Lee, J.K., et al. (2004) Enhanced transfection efficiency of PAMAM

Published by Woodhead Publishing Limited, 2013

dendrimer by surface modification with L-arginine. *J Control Release*, **99**, 445–56.
27. Kim, T.I., Bai, C.Z., Nam, K. and Park, J.S. (2009) Comparison between arginine conjugated PAMAM dendrimers with structural diversity for gene delivery systems. *J Control Release*, **136**, 132–9.
28. Arima, H., Kihara, F., Hirayama, F. and Uekama, K. (2001) Enhancement of gene expression by polyamidoamine dendrimer conjugates with alpha-, beta-, and gamma-cyclodextrins. *Bioconjug Chem*, **12**, 476–84.
29. Kihara, F., Arima, H., Tsutsumi, T., Hirayama, F. and Uekama, K. (2002) Effects of structure of polyamidoamine dendrimer on gene transfer efficiency of the dendrimer conjugate with alpha-cyclodextrin. *Bioconjug Chem*, **13**, 1211–19.
30. Kihara, F., Arima, H., Tsutsumi, T., Hirayama, F. and Uekama, K. (2003) In vitro and in vivo gene transfer by an optimized alpha-cyclodextrin conjugate with polyamidoamine dendrimer. *Bioconjug Chem*, **14**, 342–50.
31. Choi, J.S., Ko, K.S., Park, J.S., Kim, Y.H., Kim, S.W., et al. (2006) Dexamethasone conjugated poly(amidoamine) dendrimer as a gene carrier for efficient nuclear translocation. *Int J Pharm*, **320**, 171–8.
32. Ma, K., Hu, M.X., Qi, Y., Zou, J.H., Qiu, L.Y., et al. (2009) PAMAM-triamcinolone acetonide conjugate as a nucleus-targeting gene carrier for enhanced transfer activity. *Biomaterials*, **30**, 6109–18.
33. van Duijvenbode, R.C., Borkovec, M. and Koper, G. J. M. (1998) Acidbase properties of poly(propylene imine) dendrimers. *Polymer*, **39**, 2657–64.
34. Zinselmeyer, B.H., Mackay, S.P., Schatzlein, A.G. and Uchegbu, I.F. (2002) The lower-generation

Published by Woodhead Publishing Limited, 2013

polypropylenimine dendrimers are effective gene-transfer agents. *Pharm Res*, **19**, 960–7.

35. Schatzlein, A.G., Zinselmeyer, B.H., Elouzi, A., Dufes, C., Chim, Y.T., et al. (2005) Preferential liver gene expression with polypropylenimine dendrimers. *J Control Release*, **101**, 247–58.
36. Omidi, Y., Hollins, A.J., Drayton, R.M. and Akhtar, S. (2005) Polypropylenimine dendrimer-induced gene expression changes: The effect of complexation with DNA, dendrimer generation and cell type. *J Drug Target*, **13**, 431–43.
37. Kim, T.I., Baek, J.U., Zhe Bai, C. and Park, J.S. (2007) Arginine-conjugated polypropylenimine dendrimer as a non-toxic and efficient gene delivery carrier. *Biomaterials*, **28**, 2061–7.
38. Zhang, W., Chen, Z., Song, X., Si, J. and Tang, G. (2008) Low generation polypropylenimine dendrimer graft beta-cyclodextrin: an efficient vector for gene delivery system. *Technol Cancer Res Treat*, **7**, 103–8.
39. Russ, V., Günther, M., Halama, A., Ogris, M. and Wagner, E. (2008) Oligoethylenimine-grafted polypropylenimine dendrimers as degradable and biocompatible synthetic vectors for gene delivery. *J Control Release*, **132**, 131–40.
40. Hao, J., Sha, X., Tang, Y., Jiang, Y., Zhang, Z., et al. (2009) Enhanced transfection of polyplexes based on pluronic-polypropylenimine dendrimer for gene transfer. *Arch Pharm Res*, **32**, 1045–54.
41. Koppu, S., Oh, Y.J., Edrada-Ebel, R., Blatchford, D.R., Tetley, L., et al. (2010) Tumor regression after systemic administration of a novel tumor-targeted gene delivery system carrying a therapeutic plasmid DNA. *J Control Release*, **143**, 215–21.

Published by Woodhead Publishing Limited, 2013

42. Lemarie, F., Croft, D.R., Tate, R.J., Ryan, K.M. and Dufes, C. (2012) Tumor regression following intravenous administration of a tumor-targeted p73 gene delivery system. *Biomaterials*, **33**, 2701–9.
43. Klingler, J. and Kaufmann, D. (2012) Polypropylenimine generation four: a suitable vector for targeted gene alteration in vitro. *J Drug Target*, **20**, 474–80.
44. Lukasavage, W. J., Kolc, J. and Dankewalter, R.G. (1981) US Patent 4 (289):872 210.
45. Choi, J.S., Lee, E.J., Choi, Y.H., Jeong, Y.J. and Park, J.S. (1998) Poly(ethylene glycol)-block-poly(l-lysine) dendrimer: novel linear polymer/dendrimer block copolymer forming a spherical water-soluble polyionic complex with DNA. *Bioconjug Chem*, **10**, 62–5.
46. Choi, J.S., Joo, D.K., Kim, C.H., Kim, K. and Park, J.S. (2000) Synthesis of a barbell-like triblock copolymer, poly(l-lysine) dendrimer-block-poly(ethylene glycol)-block-poly(l-lysine) dendrimer, and its self-assembly with plasmid DNA. *J Am Chem Soc*, **122**, 474–80.
47. Toth, I., Sakthivel T., Wilderspin, A.F., Bayele, H., Lee, C.A., et al. (1999) Novel cationic lipidic peptide dendrimer vectors: in vitro gene delivery. *STP Pharma Sciences*, **9**, 93–9.
48. Shah, D.S., Sakthivel, T., Toth, I., Florence, A.T. and Wilderspin, A.F. (2000) DNA transfection and transfected cell viability using amphipathic asymmetric dendrimers. *Int J Pharm*, **208**, 41–8.
49. Ohsaki, M., Okuda, T., Wada, A., Hirayama, T., Niidome, T., et al. (2002) In vitro gene transfection using dendritic poly(L-lysine). *Bioconjug Chem*, **13**, 510–17.
50. Kawano, T., Okuda, T., Aoyagi, H. and Niidome, T. (2004) Long circulation of intravenously administered plasmid DNA delivered with dendritic poly(L-lysine) in the blood flow. *J Control Release*, **99**, 329–37.

Published by Woodhead Publishing Limited, 2013

51. Okuda, T., Sugiyama, A., Niidome, T. and Aoyagi, H. (2004) Characters of dendritic poly(L-lysine) analogues with the terminal lysines replaced with arginines and histidines as gene carriers in vitro. *Biomaterials*, **25**, 537–44.
52. Yamagata, M., Kawano, T., Shiba, K., Mori, T., Katayama, Y., et al. (2007) Structural advantage of dendritic poly(L-lysine) for gene delivery into cells. *Bioorg Med Chem*, **15**, 526–32.
53. Luo, K., Li, C., Wang, G., Nie, Y., He, B., et al. (2011) Peptide dendrimers as efficient and biocompatible gene delivery vectors: Synthesis and in vitro characterization. *J Control Release*, **155**, 77–87.
54. Moreno, K.X. and Simanek, E.E. (2008) Identification of diamine linkers with differing reactivity and their application in the synthesis of a melamine dendrimers. *Tetrahedron Lett*, **49**, 1152–4.
55. Steffensen, M.B. and Simanek, E.E. (2003) Chemoselective building blocks for dendrimers from relative reactivity data. *Org Lett*, **5**, 2359–61.
56. Mintzer, M.A., Merkel, O.M., Kissel, T. and Simanek, E.E. (2009) Polycationic triazine-based dendrimers: effect of peripheral groups on transfection efficiency. *New J Chem*, **33**, 1918–25.
57. Merkel, O.M., Mintzer, M.A., Sitterberg, J., Bakowsky, U., Simanek, E.E., et al. (2009) Triazine dendrimers as nonviral gene delivery systems: effects of molecular structure on biological activity. *Bioconjug Chem*, **20**, 1799–806.

Published by Woodhead Publishing Limited, 2013

14

Cyclodextrins and cyclodextrin-containing polymers

DOI: 10.1533/9781908818645.287

Abstract: Cyclodextrins are cyclic (α-1, 4)-linked oligosaccharides of α-D-glucopyranose with a central hydrophobic cavity and hydrophilic outer surface usually derived from enzymatic degradation of starch. Commonly, existing CDs include α-CD, β-CD and γ-CD which consist of six, seven and eight glucopyranose units, respectively. The CD moieties can be embedded in the polymer matrix or grafted onto polymers as pendent groups. CDs and cyclodextrin-containing polymers have been extensively investigated as gene delivery vehicles. All the polymers grafted with CDs showed reduced cytotoxicity and improved transfection efficacy.

Key words: cyclodextrins, glucopyranose, cyclodextrin-containing polymers, co-monomer, PEGylated, imidazole, oligoethyleneimine, adamantine.

14.1 Introduction

Cyclodextrins are naturally occurring cyclic (α-1,4)-linked oligosaccharides of α-D-glucopyranose containing a central hydrophobic cavity and a hydrophilic outer surface derived from enzymatic degradation of starch. Due to the absence of free rotation about the bonds that connect the glucopyranose units, the CDs are not cylindrical molecules but are toroidal or cone-shaped. The structural distribution consists of primary hydroxyl groups on the narrow side of the torus and secondary hydroxyl groups on the wider edge (Figure 14.1). Commonly existing CDs include α-CD, β-CD, and γ-CD, which consist of six, seven, and eight glucopyranose units, respectively. The most important characteristic of the cyclodextrin-containing polymers (CDPs) gene delivery system is that the polyplexes formed between the polymers and DNA can further be modified by inclusion complex formation due to the presence of large numbers of CD moieties.[1,2] The CDs have been suggested to improve the bioavailability of entrapped molecules due to their inherent membrane absorption-enhancing properties and their ability to stabilize biomolecules in the physiological milieu by shielding them from non-specific interactions.[3] Interaction of CDs with cell membranes leads to release of some membrane components (e.g. phospholipids or cholesterol) followed by destabilization and permeabilization.[4] This membrane destabilization was thought to occur due to the presence of the hydrophobic cavity inside the CDs; however, the exact mechanism could not be established. A number of cationic polymers, such as linear and branched PEI, chitosan, and PAMAM dendrimer, were modified by grafting CDs onto the polymers, and studied for gene delivery. All the CDPs showed reduced cytotoxicity due to the grafted CD moieties. The CD moieties can be embedded in the polymer matrix or grafted onto polymers as pendant groups.

Published by Woodhead Publishing Limited, 2013

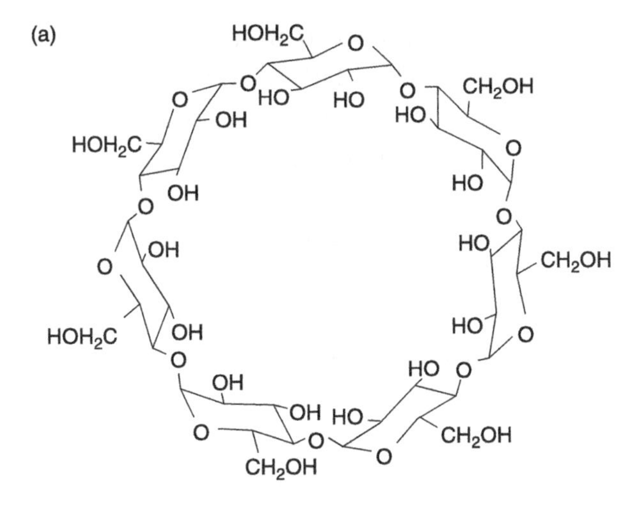

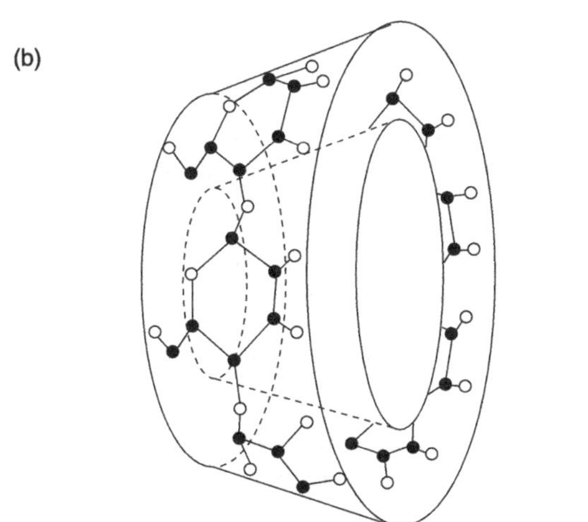

Figure 14.1 (a) The chemical structure, (b) the toroidal shape of the β-cyclodextrin molecule

14.2 Cyclodextrin-embedded polymers

The application of CDPs in gene delivery was pioneered by Davis et al., who reported the use of β-CD-containing linear cationic polymers synthesized by copolymerizing difunctionalized β-CD monomers (AA) with other

Published by Woodhead Publishing Limited, 2013

difunctionalized co-monomers (BB) such that an AABBAABB product was formed.[5] The CDP efficiently condensed pDNA (~5 kbp) to form nanoparticles of size ~100–150 nm at a N/P ratio of 1.5 (ratio of amines of the β-CDP to the phosphate of DNA) and effectively transfected cells at N/P ratios above 10. The transfection studies done in BHK-21 and CHO-K1 cells revealed higher transfection efficacies, comparable to those of PEI and Lipofectamine. Some cell line-dependent toxicities were observed for serum-free transfections, while in the presence of 10% serum toxicity was absent up to a N/P ratio of 70. Moreover, intravenous or intraperitoneal administration of high doses of β-CDP (200 mg/kg) in mice showed no mortality.[5]

The variations in co-monomer B result in changes in polymer charge density and hydrophobicity of CDP, which could further influence the biological properties of the polymer; to elucidate this, the effect of co-monomer B structure on DNA delivery and toxicity was investigated.[6] Polymers with four, five, six, seven, eight, and ten methylene units (β-CDP4, 5, 6, 7, 8, and 10) in the co-monomer B molecule condensed pDNA to form nanoparticles of size ~124–152 nm. Particle size depended on polymer and DNA concentrations in the polyplex formulations, which, however, remained unaffected above a N/P ratio of 3.[6] Transfection efficiency was dependent on co-monomer B length, with up to a 20-fold difference between polymers; maximal transfection efficiency was obtained with β-CDP6 polymer. Cytotoxicity was absent in the case of β-CDP8, while it increased for all co-monomer B lengths differing from that used in β-CDP8.[6]

Modulation of polycations by direct grafting of additional components, e.g. PEG, either before or after DNA complexation, tends to interfere with polymer/DNA binding interactions; to circumvent this, surface-PEGylated β-CDP polyplexes were formed by self-assembly of the polyplexes

Published by Woodhead Publishing Limited, 2013

with adamantane-PEG conjugates.[7] PEGylated β-CDP polyplexes were stable at physiological salt concentration (size ~110 nm), while unmodified polyplexes rapidly aggregated and precipitated in salt solutions (no discrete particles were observed). Galactosylated β-CDP-based particles revealed selective targeting to hepatocytes via the asialoglycoprotein receptor with approximately tenfold higher efficiency than glucosylated particles (control).[7]

The influence of polycation structure on cytotoxicity and gene delivery efficiency was investigated by synthesizing a series of amidine-based polycations that contained the carbohydrates D-trehalose and β-CD.[8] Polyplexes formed from all the polycations were between 80 and 100 nm in diameter. The gene delivery efficiency in BHK-21 cells showed dependence on the carbohydrate size (trehalose vs. CD) and its distance from the charge centers. The toxicity was found to increase as the charge center was removed further from the carbohydrate unit. Moreover, the toxicity was reduced with enlargement in the size of the carbohydrate moiety from trehalose to β-CD. On the other hand, removal of the carbohydrate in the polycations produced high toxicity.[8] Further, comparison between series of quaternary ammonium polycations containing N,N,N',N'-tetramethyl-1,6-hexanediamine, D-trehalose, and β-CD suggested that the quaternary ammonium analogues exhibit lower gene expression values and similar toxicities to their amidine analogues.[9] Additionally, transfection experiments conducted in the presence of chloroquine reveal increased gene expression from quaternary ammonium-containing polycations and not from their amidine analogues. This high transfection efficiency was thought to be due to the inability of quat-based polyplexes to escape from endosomes, since the addition of chloroquine to transfection experiments increased gene expression for quat-based but not amidine

Published by Woodhead Publishing Limited, 2013

analogues.[9] Furthermore, longer alkyl regions between the CD and the charge centers in the polycation backbone increased transfection efficiency and toxicity in BHK-21 cells, while increasing hydrophilicity of the spacer (alkoxy versus alkyl) results in lower toxicity.[10] Also, γ-CD-based polycations are shown to be less toxic than otherwise identical β-CD-based polycations. Later, structure–activity relationship (SAR) studies concluded that low MW polymers (~10 kDa, DP 5–8) with the CD units sufficiently distant from amidine cationic centers were the optimal architectures in terms of both high delivery efficiency and low cytotoxicity.[11]

Transfection efficiency of CDP has been shown to improve with incorporation of imidazole groups at low charge ratios (of most interest for systemic administration) without significant increase in toxicity.[11] Enhanced transfection efficiency of the imidazole-containing variant (CDPim) was attributed to increased buffering.[12] To decipher the mechanism(s) by which these two related materials exhibit differences in gene delivery, CDP and CDPim were investigated in detailed physio-chemical and biological studies.[13] The CDPim polymer displayed comparable DNA uptake to CDP and improved transfection efficiency in cultured cells, including cells with a cytoplasmic expression system. However, it remained unclear whether the improved transfection efficiency of CDPim was a result of enhanced endosomal escape, as it was also observed that CDPim generates greater amounts of unpackaged intracellular nucleic acids than CDP. Moreover, the increase in unpackaged intracellular DNA with CDPim over CDP is not consistent with cell-free measurements, and emphasizes the need to evaluate the behavior of non-viral gene delivery vectors within cellular environments.[13]

Srinivasachari et al. employed a copper (I)-catalyzed "click reaction" to synthesize β-CD copolymers with variation in

Published by Woodhead Publishing Limited, 2013

oligoethyleneamine stoichiometry – $Cd1_{46}$, $Cd2_{44}$, $Cd3_{49}$, and $Cd4_{47}$ (one to four oligoethyleneamines, respectively, in the repeat unit, and similar degrees of polymerization, n_w=44–49) – and with variation in polymer length (four ethyleneamines in the repeat unit): $Cd4_{27}$, $Cd4_{47}$, $Cd4_{93}$, and $Cd4_{200}$ (n_w=27, 47, 93, 200).[14] The polymers compacted pDNA into nanoparticles between ~75- and 120 nm. It was observed that within Series 1 the polyplex size decreased as the amine density increased, while in Series 2 the polyplex size tended to increase with the polymer length, with the exception of $Cd4_{47}$. DNAse degradation assay revealed that these polymers were quite capable of protecting pDNA from nuclease degradation, irrespective of the degree of polymerization, and the pDNA protection capability increased with amine stoichiometry in the polymeric structure. The gene expression profiles of the resulting "click" polymers in HeLa cells were mostly dependent on the oligoethyleneamine/CD ratio, with the polymers having longer oligoethyleneamine chains showing highest efficacies.[14]

Two novel biodegradable polymers of LMW PEI (MW 600 Da) cross-linked by (2-hydroxypropyl)-β-cyclodextrin (HP-β-CDs) or (2-hydroxypropyl)-γ-cyclodextrin (HP-γ-CD) were developed to improve transfection efficiency.[15] The polymers effectively condensed pDNA, resulting in nanoparticles of size less than 300 nm. The newly synthesized polymers exhibited significantly lower cytotoxicity as compared with PEI 25 kDa in SKOV-3 cells. Further, the transfection efficiency of polymer/DNA nanoparticles was 1.5–1.7-fold higher than that of PEI 25 kDa/DNA complexes and over 20-fold higher than that of PEI 600 Da/DNA complexes.[15]

In another study, 2-hydroxypropyl-α-cyclodextrin (HP-α-CD) was conjugated to LMW PEI (MW 600 Da) by reaction with carbonyl diimidazole (CDI).[16] The carbamate

Published by Woodhead Publishing Limited, 2013

linkages ensure polymer biodegradability, which contributes to reducing toxicity. The particle size of the complexes formed with HP-α-CD-PEI/DNA complexes was ~186 nm, and that of the PEI/DNA complexes was ~4800 nm. The cell viability was markedly higher for HP-α-CD-PEI (~95% at a concentration of 1250 nmol/ml) in contrast to PEI 25 kDa (~19% at concentration of 300 nmol/ml). Transfection efficiency of HP-α-CD-PEI/DNA complexes was up to 5.5-fold higher as compared with PEI (25 kDa) polyplexes in SKBR-3 human breast cancer cells in complete media.[16] Further, the transfection efficiency of the HP-α-CD was comparable to, and at higher polymer/DNA ratios even higher than, that offered by PEI 25 kDa in neural cells.[17] Moreover, intrathecal injection of HP-α-CD-PEI/DNA complexes into the rat spinal cord yielded gene expression levels close to that obtained by PEI 25 kDa.[17]

To develop a tumor-targeted gene delivery vector, folate was conjugated to low MW PEI (600 Da)-grafted CD to form polymer H_1.[18] The polymer H_1 condensed pDNA at N/P ratios between 5 and 30 to form particles of size less than 120 nm and zeta potentials less than +10 mV. The *in vitro* transfection efficiency of H_1 was more than 50% in various tumor cell lines (U138, U87, B16, and Lovo), which was improved by the presence of FBS or albumin in the transfection medium. Also, cell viability was significantly higher for H_1 than for high MW PEI 25 kDa.[18] Additionally, *in vivo* optical imaging showed that the efficiency of H_1-mediated transfection (50 μg luciferase plasmid (pLuc), N/P ratio = 20/1) was comparable to that of adenovirus-mediated luciferase transduction (1×10^9 pfu) in melanoma-bearing mice, and it did not induce any toxicity in the tumor tissue.[18] Later, HP-γ-CD grafted with PEI (600 Da) was coupled to MC-10 oligopeptide containing a sequence of Met-Ala-Arg-Ala-Lys-Glu to target the human epidermal

Published by Woodhead Publishing Limited, 2013

growth factor receptor 2, which is over-expressed in many breast and ovarian cancers.[19] The particle size of HP-γ-CD-PEI/DNA complexes at a N/P ratio of 40 was ~170–200 nm, with zeta potential of about +20 mV. Polymer HP-γ-CD-PEI exhibited very low cytotoxicity, strong targeting specificity to the HER2 receptor, and high efficiency in DNA delivery to target cells both *in vitro* and *in vivo*. The delivery and the therapeutic efficiency of IFN-α gene mediated by the new gene vector were also studied in a mouse model. The HP-γ-CD-PEI-mediated IFN-α gene delivery showed significantly enhanced anti-tumor effect in tumor-bearing nude mice as compared with PEI (25 kDa), HP γ-CD-PEI, and other controls, suggesting the therapeutic potential of this vector.[19]

The condensation of DNA by β-CD, an amphiphilic cationic connector (DC-Chol or adamantane derivative Ada2), was evaluated by using surface-enhanced Raman spectroscopy (SERS).[20] The amount of cationic connector in the medium was used to control the charge of the polymeric vector. The SERS spectra showed a decrease of signal intensity when the vector/DNA charge ratio (Z+/–) increased. At the highest ratio (Z+/– = 10) the signal was sixfold and threefold less intense than the DNA reference signal for Ada2 and DC-Chol polyplexes, respectively, suggesting that adenyl residues have a reduced accessibility as DNA is bound to the vector.[20] The study demonstrated that the cationic charges neutralizing the negative charges of DNA result in the formation of stable polyplexes. *In vitro* transfection efficiency results exhibited a twofold increase with the ternary complex as compared with β-CD.[20] Another study employed viscometry and small angle neutron scattering (SANS) techniques to investigate the formation of ternary complex between β-CD, cationic surfactant *n*-dodecyltrimethylammonium chloride (DTAC), and low MW DNA (herring sperm DNA fragments).[21] The factors

Published by Woodhead Publishing Limited, 2013

that determine the poly(β-CD)/DTAC/DNA complexation in aqueous solution include reversible inclusion interactions between the CD cavities of β-CD and the alkyl group of DTAC, leading to the formation of a polycation, and electrostatic interactions between the opposite charges of the cationic surfactant and anionic DNA. Also, the β-CD/DTAC/DNA complexes were reported to be completely reversible in the presence of 150 mM of NaCl salt.[21]

A distinct advantage of CDPs is that the nanoparticles resulting from complexation with ODNs/pDNA can be modulated at their surface by exploiting the intrinsic CD inclusion capabilities. For instance, CDP/pDNA complexes were modified by inclusion of the adamantane moiety of adamantane-modified PEG (Ad-PEG) into CD cavities, which provided a steric corona around the particles that prevented aggregation and non-specific interactions with biological components.[22] The size of CDP/pDNA nanoparticles was ~100 nm, which steadily increased upon addition of PBS, suggesting aggregation, while that of PEGylated CDP/pDNA complexes was ~110 nm and remained unaltered upon addition of PBS. CDP/pDNA polyplexes exhibited a significant two to threefold higher luciferase expression as compared with naked DNA, whereas PEGylation of polyplexes reduced the observed luciferase activity nearly to the level achieved with naked pDNA. Further, polyCDplexes coated with galactosylated Ad-PEG were shown to exhibit selectivity towards hepatocytes with galactose-specific membrane receptors.[7] Also, polyCDplexes shielded with transferrin-modified Ad-PEG (Ad-PEG-TF) transfected luciferase-encoding gene to K562 human myelogeneous leukemia cells with better efficiency than non-targeted particles.[23,24]

A series of cationic star polymers were synthesized by conjugating multiple OEI arms onto α-CD core as non-viral

Published by Woodhead Publishing Limited, 2013

gene delivery vectors.[25] The molecular structures of the α-CD-OEI star polymers, which contained linear or branched OEI arms with different chain lengths ranging from 1 to 14 ethylenimine units, were characterized by using size exclusion chromatography and nuclear magnetic resonance (NMR) techniques. The α-CD-OEI star polymers were studied in terms of their DNA binding capability, formation of nanoparticles with pDNA, cytotoxicity, and gene transfection in cultured cells. All the α-CD-OEI star polymers could inhibit the migration of pDNA on agarose gel via formation of complexes with pDNA, and the complexes formed nanoparticles with sizes ranging from 100 to 200 nm at N/P ratios of 8 or higher. The star polymers displayed much lower *in vitro* cytotoxicity than that of branched PEI of MW 25 kDa. The α-CD-OEI star polymers showed excellent gene transfection efficiency in HEK 293 and COS-7 cells. Generally, the transfection efficiency increased with an increase in the OEI arm length. The star polymers with longer and branched OEI arms showed higher transfection efficiency.

14.3 Polymers with cyclodextrins as pendant groups

The properties of pre-existing polymers employed in gene delivery have been modulated by attachment of CDs. Since most of the polymers used for gene delivery were cationic in nature, they could be manipulated to achieve improved transfection efficiency, to avoid or diminish their toxicity, or to provide them with additional capabilities. To incorporate the advantages of CDs, i.e. low toxicity and the ability to modify the polyplex via inclusion complex formation, high MW LPEI and BPEI (25 kDa) were modified by reacting with

Published by Woodhead Publishing Limited, 2013

6-mono-tosyl-β-cyclodextrin.[26] CD grafting resulted in decrease in transfection efficiency; 10% amine modification showed a 2-fold decrease in luciferase activity and 16% amine modification reduced expression 4-fold as compared with unmodified PEI. However, cell viability was increased with the increase in grafting in PC3 cells; high levels of cyclodextrin grafting increase IC_{50} values by over 20-fold (4.8 mM for 12% grafting and 6.7 mM for 16% grafting). CD-BPEI and CD-LPEI polymers self-assembled with and condensed pDNA to small spherical nanoparticles with size ~100 to 160 nm. PEGylation with Ad-PEG at 1:1 Ad:CD ratio reduced aggregation rate by forming a protective hydrophilic PEG brush layer on the particle surface.[26] Intravenous administration of up to 120 μg of DNA formulated with CD-LPEI + AD-PEG did not induce any acute toxicity, while LPEI-DNA polyplexes at this dose were observed to be lethal to mice. Furthermore, PEGylated CD-LPEI/DNA polyplexes showed reduced aggregation and primarily accumulated in liver in contrast to lung (4-fold higher amount in liver than in lung) and exhibited gene expression localized to liver only.[26] The CD/DNA polyplexes were further derivatized with targeting ligands using their inherent molecular inclusion capabilities. PEI with ~10% grafting with β-CD (CD-PEI) was modulated with hydrophobic palmitate group (pal-HI)-modified human insulin.[27] CD-PEI efficiently condensed DNA, exhibited 4-fold higher gene expression in HEK293 cells as compared with unmodified PEI, and was almost non-toxic. Also, addition of the pal-HI to CD-PEI enhanced gene expression by more than an order of magnitude compared with unmodified PEI, either with or without the pal-HI.[27]

CD grafting to PEI was also utilized to promote immobilization of PEI/DNA nanoparticles onto solid

Published by Woodhead Publishing Limited, 2013

surfaces. CD-PEI nanoparticles were immobilized on Ad-modified self-assembled monolayers for controlled surface-mediated gene delivery.[28] Surface plasmon resonance studies suggested that CD-PEI nanoparticles were specifically immobilized on the chip surface by CD-Ad inclusion complex formation, while minimal adsorption was detected with PEI-based nanoparticles. Binding affinity of CD-PEI nanoparticles to the adamantane surfaces was several orders of magnitude higher than binding of single cyclodextrin molecules due to multivalent interactions.[28] Later, a CD-based polymeric assembly was designed to act as a dual (gene and drug) delivery system. Inclusion-driven supramolecular assembly of a β-CD-grafted PEI with poly(β-benzyl-L-aspartate) (PBLA) generated core–shell structured nanoparticles, with a hydrophobic core that can locate hydrophobic drugs, and a cationic shell intended to condense DNA.[29] Further, the transfection efficiency towards osteoblast cells using particles with and without drug loading (dexamethasone, DMS) exhibited less efficient transfection than PEI-based polyplexes; also, the drug loading exerted a mild positive effect on both cell viability and gene transfer.[29]

CD was also grafted to poly(ε-lysine) to form a sunflower-shaped β-CD-conjugated poly(ε-lysine) (β-CDPL) polyplex for gene delivery.[30] The confocal microscopy data suggested that β-CDPL polyplexes are efficiently internalized even at short incubation times; after 0.5 h of transfection, significant cellular uptake was observed, while PL polyplexes showed limited internalization after 3.0 h. Further, the β-CDPL polyplexes at N/P ratios of 5 and 10 exhibited transfection efficiencies fourfold and tenfold higher than those of PL and LPEI, respectively. It was reasoned that the β-CDs facing the outside of the polyplex promoted the removal of cholesterol from the cell membrane, introducing local membrane disturbances and facilitating uptake of complexes by

Published by Woodhead Publishing Limited, 2013

endocytosis, and the secondary amines of β-CDPL utilized the proton-sponge effect to produce significantly enhanced transfection.[30]

CD was grafted to chitosans (MW 110 and 10 kDa) to yield hybrid chitosan/CD nanoparticles which were then loaded with pSEAP (plasmid DNA model that encodes the expression of secreted alkaline phosphatase) to form nanoparticles of size ~100–200 nm, depending on chitosan MW, and a positive surface charge of ~+22 to +35 mV.[31] Chitosan/CD nanoparticles exhibited a significantly lower cytotoxicity in Calu-3 cells than those composed of solely chitosan (CS); IC_{50} values were threefold higher for CD-containing nanoparticles than for the CS control nanoparticles. Cellular uptake studies showed that the nanoparticles were effectively internalized by the cells, and localized around the cell nuclei. The transfection efficiency of the different formulations, measured by the concentration of secreted gene product (SEAP), indicated that all the nanoparticles were able to elicit a significantly higher response than the naked DNA (control), the transfection efficiency being more important for low MW chitosan nanoparticles than for those composed of medium MW chitosan.[31] Two water-soluble chitosan-graft-(PEI-β-CD) (CPC) cationic copolymers were synthesized via reductive amination between oxidized chitosan and low MW β-CD-PEI.[32] Gene transfection experiments showed improved performance mediated by both the polymers as compared with native chitosan in HEK293, L929, and COS7 cell lines. Further investigation of the gene transfection mediated by CPC2/DNA complexes showed both time and dose-dependency in the tested cell lines, where the polymer showed higher level luciferase expression than commercially available branched PEI (25 kDa) under conditions of high dose or extended time.[32]

14.4 Cyclodextrins as adjuvants for enhanced gene delivery

The addition of cyclodextrin derivatives as formulation excipients has been reported to improve the transfection efficiency of various gene delivery vectors. One of the initial studies reported a sixfold improvement in gene expression in rat lung upon addition of β-CD (1%) to the original DNA:lipid formulations.[33] The enhanced transfection efficacy was attributed to CD membrane permeation enhancement capabilities in this tissue. Addition of methyl-β-cyclodextrin solubilized cholesterol (MBC) to DOTAP and Superfect further improved their transfection efficiency by 3.8-fold and 2.6-fold, respectively.[34] The gene expression of β-galactosidase was easily detectable within 1 h of transfection and reached a maximal value at 48 h. Intracellular localization studies suggested that MBC-derivatized DOTAP/DNA complexes were found in both the nucleus and the cytoplasm. Further, intravesical administration of MBC-derivatized DOTAP/DNA complexes to mice bladders resulted in enhanced gene expression after 2 days, which remained confined to the bladder and was sustained for up to 30 days after transfection.[34]

Another study reported inclusion of β-CD into PAMAM dendrimer/DNA formulations to form dendrimer/DNA/β-CD complexes.[35] The size of the complexes formed with 50 μg/ml of DNA at a N/P ratio of 1 decreased from 156 nm to 5.8–21.2 nm in 0.025–0.1% w/vol β-CDs. Also, sulfonated β-CDs bind to dendrimers, and high concentrations may displace DNA in the dendrimer/DNA complex. Further, *in vitro* chloramphenicol acetyltransferase (CAT) expression on the surface of collagen membranes increased ~200–fold when mediated by dendrimer/DNA/β-CD formulations as compared with formulations without β-CD. Maximal

Published by Woodhead Publishing Limited, 2013

transfection efficiency was obtained with sulfated anionic CDs, which facilitated gradual dissociation of cationic polymer–DNA complexes.[35]

14.5 Cyclodextrin-based polyrotaxanes

Rotaxanes are unique supramolecular structures in which a cyclic molecule is threaded onto an “axle” molecule and end-capped by bulky groups at the terminal of the “axle” molecule. Polyrotaxanes and polypseudorotaxanes are polymers with multiple cyclic molecules threaded onto a polymer chain with or without bulky end caps. Due to their low cytotoxicity, controlled size, and unique architecture, CD-based polyrotaxanes and polypseudorotaxanes have been exploited as gene delivery vectors. Polyrotaxanes with cationic polymer-modified CD rings showed good DNA binding ability, low cytotoxicity, and high gene transfection efficiency.[36–38]

One of the first studies reported formation of polyrotaxanes by threading α-CD molecules over PEG and poly(ε-caprolactone) (PCL) chains of ternary block copolymers of PEG, PCL, and PEI.[38] Due to reduction in hydrophobic interaction between PCL blocks, the resulting polyrotaxanes displayed an enhanced solubility in comparison with the ternary block copolymers. Ethidium bromide fluorescence quenching assays demonstrated that polyrotaxanes complexed DNA as efficiently as BPEI 25 kDa. The size of the resultant polyrotaxane/DNA polyplexes was observed to be ~ 200 nm, with neutral surface charge and efficient cellular uptake. Although the transfection efficiency was similar to PEI in mouse embryonic fibroblast 3T3 cells, toxicity was 100-fold lower, allowing the administration of N/P ratios of up to 20.[38] The virtually neutral surface potential (z potential)

Published by Woodhead Publishing Limited, 2013

of these complexes did not impair efficient cell uptake and could account for the absence of toxic effects.

Three different cationic polyrotaxanes with different lengths of OEI were synthesized by grafting of OEI to β-CD threaded on a pluronic PEO-PPO-PEO triblock copolymer chain and end-capped with 2,4,6-trinitrobenzene sulfonate.[37] The cationic supramolecular gene delivery vectors showed good DNA binding ability, low cytotoxicity, and high gene transfection efficacy, similar to PEI (25 kDa), at the optimized N/P ratio and MW. They even exhibited higher transfection efficiency than dimethylaminoethyl (DMAEC)-α-CD polyrotaxane systems. Further, LPEI-γ-CD polypseudorotaxanes were developed as a safer gene carrier alternative.[36] Although these supramolecular constructs condensed pDNA less efficiently than naked LPEI (22 kDa), their far lower toxicity and the better cellular uptake of the corresponding rotaplexes fully compensated for this, achieving similar gene expression levels to PEI-based polyplexes in NIH3T3 cells. The cellular uptake of pDNA in the LPEI/γ-CD polyplex was enhanced by free γ-CDs released from the polyplex, which might accelerate the cellular uptake through enhanced membrane affinity.

14.6 Conclusions

Cyclodextrins improve the gene delivery efficacy of pre-existing cationic polymers or dendrimers, due to their capacity to permeabilize cell membranes by affecting cholesterol distribution. However, the efficiency of the best cyclodextrin-based gene delivery systems currently developed remains several orders of magnitude poorer as compared with viral vectors. Nonetheless, the possible combination of covalent and supramolecular approaches offers exciting

Published by Woodhead Publishing Limited, 2013

avenues for the fabrication of custom-made vectors. The CD polymer-based vector CALAA-01 has already entered clinical trials ten years after its conceptualization. It is expected that, due to development in understanding of CD-based gene delivery mechanisms, more dendrimer-based vectors will enter clinical trials soon.

14.7 References

1. Davis, M.E. and Brewster, M.E. (2004) Cyclodextrin-based pharmaceutics: past, present and future. *Nat Rev Drug Discov*, **3**, 1023–35.
2. Pack, D.W., Hoffman, A.S., Pun, S. and Stayton, P.S. (2005) Design and development of polymers for gene delivery. *Nat Rev Drug Discov*, **4**, 581–93.
3. Doziuk, H. (ed.) (2006) *Cyclodextrins and their complexes*. Wiley-VCH, Weinheim.
4. Zidovetzki, R. and Levitan, I. (2007) Use of cyclodextrins to manipulate plasma membrane cholesterol content: evidence, misconceptions and control strategies. *Biochim Biophys Acta*, **1768**, 1311–24.
5. Gonzalez, H., Hwang, S.J. and Davis, M.E. (1999) New class of polymers for the delivery of macromolecular therapeutics. *Bioconjug Chem*, **10**, 1068–74.
6. Hwang, S.J., Bellocq, N.C. and Davis, M.E. (2001) Effects of structure of beta-cyclodextrin-containing polymers on gene delivery. *Bioconjug Chem*, **12**, 280–90.
7. Pun, S.H. and Davis, M.E. (2002) Development of a nonviral gene delivery vehicle for systemic application. *Bioconjug Chem*, **13**, 630–9.
8. Reineke, T.M. and Davis, M.E. (2002) Structural effects of carbohydrate-containing polycations on gene

Published by Woodhead Publishing Limited, 2013

delivery. 1. Carbohydrate size and its distance from charge centers. *Bioconjug Chem*, **14**, 247–54.

9. Reineke, T.M. and Davis, M.E. (2002) Structural effects of carbohydrate-containing polycations on gene delivery. 2. Charge center type. *Bioconjug Chem*, **14**, 255–61.
10. Popielarski, S.R., Mishra, S. and Davis, M.E. (2003) Structural effects of carbohydrate-containing polycations on gene delivery. 3. Cyclodextrin type and functionalization. *Bioconjug Chem*, **14**, 672–8.
11. Davis, M.E., Pun, S.H., Bellocq, N.C., Reineke, T.M., Popielarski, S.R., et al. (2004) Self-assembling nucleic acid delivery vehicles via linear, water-soluble, cyclodextrin-containing polymers. *Curr Med Chem*, **11**, 179–97.
12. Kulkarni, R.P., Mishra, S., Fraser, S.E. and Davis, M.E. (2005) Single cell kinetics of intracellular, nonviral, nucleic acid delivery vehicle acidification and trafficking. *Bioconjug Chem*, **16**, 986–94.
13. Mishra, S., Heidel, J.D., Webster, P. and Davis, M.E. (2006) Imidazole groups on a linear, cyclodextrin-containing polycation produce enhanced gene delivery via multiple processes. *J Control Release*, **116**, 179–91.
14. Srinivasachari, S. and Reineke, T.M. (2009) Versatile supramolecular pDNA vehicles via "click polymerization" of β-cyclodextrin with oligoethyleneamines. *Biomaterials*, **30**, 928–38.
15. Huang, H., Tang, G., Wang, Q., Li, D., Shen, F., et al. (2006) Two novel non-viral gene delivery vectors: low molecular weight polyethylenimine cross-linked by (2-hydroxypropyl)-beta-cyclodextrin or (2-hydroxypropyl)-gamma-cyclodextrin. *Chem Commun (Camb)*, 2382–4.

Published by Woodhead Publishing Limited, 2013

16. Huang, H., Yu, H., Li, D., Liu, Y., Shen, F., et al. (2008) A novel co-polymer based on hydroxypropyl alpha-cyclodextrin conjugated to low molecular weight polyethylenimine as an in vitro gene delivery vector. *Int J Mol Sci*, **9**, 2278–89.
17. Tang, G.P., Guo, H.Y., Alexis, F., Wang, X., Zeng, S., et al. (2006) Low molecular weight polyethylenimines linked by beta-cyclodextrin for gene transfer into the nervous system. *J Gene Med*, **8**, 736–44.
18. Yao, H., Ng, S.S., Tucker, W.O., Tsang, Y.-K.-T., Man, K., et al. (2009) The gene transfection efficiency of a folate–PEI600–cyclodextrin nanopolymer. *Biomaterials*, **30**, 5793–803.
19. Huang, H., Yu, H., Tang, G., Wang, Q. and Li, J. (2010) Low molecular weight polyethylenimine cross-linked by 2-hydroxypropyl-gamma-cyclodextrin coupled to peptide targeting HER2 as a gene delivery vector. *Biomaterials*, **31**, 1830–8.
20. Burckbuchler, V., Wintgens, V., Lecomte, S., Percot, A., Leborgne, C., et al. (2006) DNA compaction into new DNA vectors based on cyclodextrin polymer: surface enhanced Raman spectroscopy characterization. *Biopolymers*, **81**, 360–70.
21. Galant, C., Amiel, C. and Auvray, L. (2005) Ternary complex formation in aqueous solution between a beta-cyclodextrin polymer, a cationic surfactant and DNA. *Macromol Biosci*, **5**, 1057–65.
22. Mishra, S., Webster, P. and Davis, M.E. (2004) PEGylation significantly affects cellular uptake and intracellular trafficking of non-viral gene delivery particles. *Eur J Cell Biol*, **83**, 97–111.
23. Bellocq, N.C., Pun, S.H., Jensen, G.S. and Davis, M.E. (2003) Transferrin-containing, cyclodextrin polymer-

Published by Woodhead Publishing Limited, 2013

based particles for tumor-targeted gene delivery. *Bioconjug Chem*, **14**, 1122–32.

24. Bartlett, D.W. and Davis, M.E. (2007) Physicochemical and biological characterization of targeted, nucleic acid-containing nanoparticles. *Bioconjug Chem*, **18**, 456–68.
25. Yang, C., Li, H., Goh, S.H. and Li, J. (2007) Cationic star polymers consisting of alpha-cyclodextrin core and oligoethylenimine arms as nonviral gene delivery vectors. *Biomaterials*, **28**, 3245–54.
26. Pun, S.H., Bellocq, N.C., Liu, A., Jensen, G., Machemer, T., et al. (2004) Cyclodextrin-modified polyethylenimine polymers for gene delivery. *Bioconjug Chem*, **15**, 831–40.
27. Forrest, M.L., Gabrielson, N. and Pack, D.W. (2005) Cyclodextrin-polyethylenimine conjugates for targeted in vitro gene delivery. *Biotechnol Bioeng*, **89**, 416–23.
28. Park, I.K., von Recum, H.A., Jiang, S. and Pun, S.H. (2006) Supramolecular assembly of cyclodextrin-based nanoparticles on solid surfaces for gene delivery. *Langmuir*, **22**, 8478–84.
29. Zhang, J., Sun, H. and Ma, P.X. (2010) Host-guest interaction mediated polymeric assemblies: multifunctional nanoparticles for drug and gene delivery. *ACS Nano*, **4**, 1049–59.
30. Choi, H.S., Yamashita, A., Ooya, T., Yui, N., Akita, H., et al. (2005) Sunflower-shaped cyclodextrin-conjugated poly(ε-Lysine) polyplex as a controlled intracellular trafficking device. *Chembiochem*, **6**, 1986–90.
31. Teijeiro-Osorio, D., Remunan-Lopez, C. and Alonso, M.J. (2009) Chitosan/cyclodextrin nanoparticles can efficiently transfect the airway epithelium in vitro. *Eur J Pharm Biopharm*, **71**, 257–63.
32. Ping, Y., Liu, C., Zhang, Z., Liu, K.L., Chen, J., et al. (2011) Chitosan-graft-(PEI-beta-cyclodextrin) copolymers

and their supramolecular PEGylation for DNA and siRNA delivery. *Biomaterials*, **32**, 8328–41.

33. Freeman, D. and Niven, R. (1996) The influence of sodium glycocholate and other additives on the in vivo transfection of plasmid DNA in the lungs. *Pharm Res*, **13**, 202–9.
34. Lawrencia, C., Mahendran, R. and Esuvaranathan, K. (2001) Transfection of urothelial cells using methyl-beta-cyclodextrin solubilized cholesterol and Dotap. *Gene Ther*, **8**, 76–8.
35. Roessler, B.J., Bielinska, A.U., Janczak, K., Lee, I. and Baker Jr, J.R. (2001) Substituted β-cyclodextrins interact with PAMAM dendrimer–DNA complexes and modify transfection efficiency. *Biochem Biophys Res Commun*, **283**, 124–9.
36. Yamashita, A., Choi, H.S., Ooya, T., Yui, N., Akita, H., et al. (2006) Improved cell viability of linear polyethylenimine through gamma-cyclodextrin inclusion for effective gene delivery. *Chembiochem*, **7**, 297–302.
37. Li, J., Yang, C., Li, H., Wang, X., Goh, S.H., et al. (2006) Cationic supramolecules composed of multiple oligoethylenimine-grafted β-cyclodextrins threaded on a polymer chain for efficient gene delivery. *Adv Mater*, **18**, 2969–74.
38. Shuai, X., Merdan, T., Unger, F. and Kissel, T. (2005) Supramolecular gene delivery vectors showing enhanced transgene expression and good biocompatibility. *Bioconjug Chem*, **16**, 322–9.

Published by Woodhead Publishing Limited, 2013

15

Poly(D,L-lactide-co-glycolide)-based nanoparticles

DOI: 10.1533/9781908818645.309

Abstract: PLGA is a copolymer of polylactic acid and polyglycolic acid; synthesized in a wide range of molecular weight by ring-opening polymerization of cyclic dimers, i.e. lactide and glycolide, in the presence of metal catalysts. Different forms of PLGA are available, depending on the ratio of lactide to glycolide used for the copolymerization. PLGA breaks down into body metabolites, i.e. lactic and glycolic acid, by hydrolysis of ester bonds, which are removed by Kreb cycle. Moreover, PLGA have been approved by U.S. Food and Drug Administration and European Medicine Agency for human use. PLGA or PLGA-based nanoparticles have been widely employed for *in vitro* and *in vivo* gene delivery, as they are biocompatible and biodegradable.

Key words: PLGA nanoparticles, polylactic acid, polyglycolic acid, non-condensing agents, copolymer, bioadhesive.

15.1 Introduction

Cationic polymers (i.e. nucleic acid condensing agents) suffer from some disadvantages, such as toxicity due to excess

Published by Woodhead Publishing Limited, 2013

positive charge, rapid clearance by the RES, poor endosomal escape, and inefficient decondensation, due to strong electrostatic interactions resulting in release of a low amount of DNA.[1] Polymers capable of forming nano-sized (i.e. non-condensing agents) vectors with either a neutral or a net negative charge have been proposed for tissue and cell-specific delivery; these possess efficient transfection efficiency with improved cell viability. DNA can either be entrapped into the polymeric matrix or adsorbed on the nanoparticle surface by electrostatic interaction by using appropriate surfactants or adding cationic polymers to the matrix (Figure 15.1).[2,3] Encapsulation provides protection against nucleases and other plasma proteins during travel through the bloodstream to the target site. Further, non-condensing vectors are not recognized by the RES, due to absence of positive charge, and thus maintain prolonged systemic circulation.[4,5]

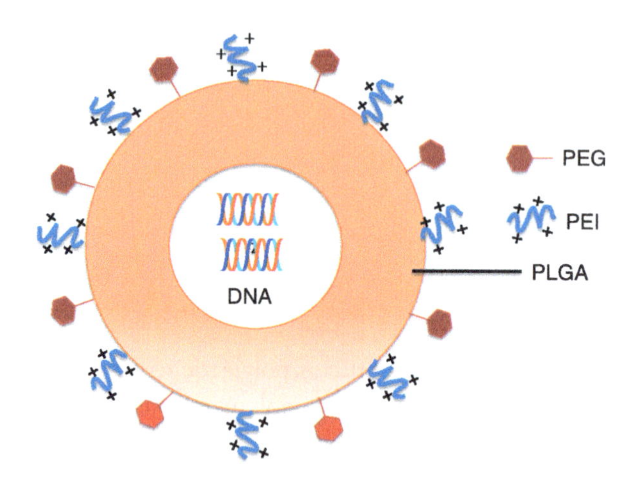

Figure 15.1 Schematic representation of PLGA nanoparticles. The surface of nanoparticles is modulated by PEG and positive charge is introduced by a cationic polymer such as PEI or chitosan

Published by Woodhead Publishing Limited, 2013

Among various non-condensing agents, PLGA has been reported as a potential candidate for gene delivery. Studies have proposed that PLGA nanoparticles are efficiently and rapidly internalized through a combination of specific and non-specific endocytic mechanisms.[6] Upon internalization, a significant amount of nanoparticles escapes the lysosomal compartment and reaches the cytosol. Surface charge reversal of PLGA nanoparticles from anionic to cationic in the acidic pH of the lysosomes is regarded as an important step that enables PLGA nanoparticles to escape the lysosomal compartment.[6,7] Following lysosomal escape, PLGA nanoparticles have been observed to remain in the cellular cytoplasm for more than 2 weeks, thereby slowly releasing their payload.[8] PLGA is broken down by hydrolysis of ester bonds into body metabolites, i.e. lactic and glycolic acids, which are removed by the Krebs cycle. Hence, the nanoparticles exhibit high biocompatibility and are nearly non-toxic. Moreover, PLGA has been approved by the US Food and Drug Administration for human use. PLGA is a copolymer of polylactic acid and polyglycolic acid, synthesized in a wide range of MWs by ring-opening polymerization of cyclic dimers, i.e. lactide and glycolide, in the presence of metal catalysts (Figure 15.2). Several modifications of PLGA have been reported for DNA delivery, especially by copolymerization with other monomers. Depending on the ratio of lactide to glycolide used for the copolymerization, different forms of PLGA are available;

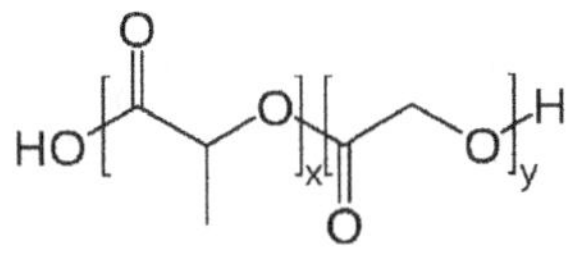

Figure 15.2 Chemical structure of PLGA. x, number of lactic acid units; y, number of glycolic acid units

Published by Woodhead Publishing Limited, 2013

these are usually identified on the basis of the ratio of the monomers used (e.g. PLGA 75:25 is a copolymer whose composition is 75% lactic acid and 25% glycolic acid). All PLGAs are amorphous and show a glass transition temperature in the range of 40–60°C.

15.2 PLGA nanoparticles for gene delivery

In one of the initial studies, PLGA polymer was used to encapsulate pDNA (alkaline phosphatase (AP), a reporter gene), forming nanoparticles of size ~600 nm with high encapsulation efficiency (70%).[9] The pDNA entrapped in nanoparticles maintained its structural and functional integrity; *in vitro* transfection resulted in significantly higher expression levels in comparison with naked pDNA. Further, AP levels increased with the increase in the transfection time, suggesting sustained release of pDNA. Increased AP expression was seen in rats after 7 and 28 days following intramuscular administration of nanoparticles, indicating sustained activity of the AP.[9]

Another seminal study reported PLGA nanoparticles of ~70 nm encapsulating 6-coumarin, with an average zeta potential of ~12.5 mV at pH 7 in 0.001 M HEPES buffer.[7] At 10 min post-incubation, nanoparticles were localized in the cytoplasmic compartment of PC3 (prostate cancer) cells. The mechanism of rapid escape was attributed to selective reversal of the surface charge of nanoparticles (from anionic to cationic) in the acidic endolysosomal compartment, which facilitates the interaction of nanoparticles with the endolysosomal membrane and escape into the cytosol. After their escape, nanoparticles delivered their payload in the cytoplasm at a slow rate,

Published by Woodhead Publishing Limited, 2013

resulting in dose-dependent transfection levels that increased with the time of transfection, suggesting sustained release of pDNA.[7]

In order to develop a controlled release delivery system for AS-ODNs, PLGA nanoparticles of size ~300 nm were prepared that encapsulated platelet-derived growth factor β-receptor antisense (PDGFβR-AS) with high encapsulation efficiency (81%) and investigated *in vitro* and *in vivo* in a balloon-injured rat restenosis model.[10] The extent of mean neointimal formation 14 days after injection of AS-nanoparticles, measured as a percentage of luminal stenosis, was $32.21 \pm 4.75\%$, in comparison with 54.89 ± 8.84 and $53.84 \pm 5.58\%$ in the blank-nanoparticles and scrambled-nanoparticles groups, respectively.[10]

Fabrication of PLGA nanoparticles employing a double emulsion solvent evaporation technique generally leads to the formation of nanoparticles with heterogeneous size distribution. Transfection efficiency of different size fractions of nanoparticles was investigated in cells by separating fractions using membrane filtration (100 nm size cut-off). The smaller-sized nanoparticle fraction (~ 70 nm) produced 27-fold higher transfection in COS-7 cells and fourfold higher transfection in HEK 293 cells for the same dose of nanoparticles as compared with larger-sized nanoparticles (~200 nm).[11] The higher transfection efficacy of smaller-sized nanoparticles was hypothesized to be due to the difference in the total number of particles present in the cells for the same mass of nanoparticles being taken up by the cells. Further, on the basis of the mean particle diameter of the two fractions of nanoparticles and the cellular uptake data, the number of smaller-sized nanoparticles taken up by the cells was estimated to be ~20-fold greater than for larger-sized nanoparticles (350×10^8 vs. 17×10^8 per mg cell protein for smaller- and larger-sized nanoparticles, respectively).[12]

Published by Woodhead Publishing Limited, 2013

To optimize the efficacy of the encapsulated therapeutic agent, the intracellular distribution as well as the tissue uptake of nanoparticles was studied.[13] Towards this objective, PLGA nanoparticles containing 6-coumarin as a fluorescent marker and osmium tetroxide as an electron microscopic marker with bovine serum albumin (BSA) as a model protein were formulated. Comparison of different physico-chemical properties of marker-loaded nanoparticles, such as particle size, zeta potential, residual PVA content, and protein-loading, with those of unloaded nanoparticles revealed no significant difference. Moreover, marker-loaded nanoparticle formulations were as non-toxic to the cells as unloaded nanoparticles. Nanoparticles loaded with 6-coumarin were found to be useful for studying intracellular nanoparticle uptake and distribution using confocal microscopy, while osmium tetroxide-loaded nanoparticles were found to be useful for studying nanoparticle uptake and distribution in cells and tissue using TEM.[13]

The anti-proliferative activity of wild-type (wt) p53 gene-loaded PLGA nanoparticles was investigated in a breast cancer cell line.[14] Nanoparticles of size ~280 nm and DNA entrapment efficiency of ~63% were formulated using a multiple-emulsion-solvent evaporation technique. Cells transfected with wt-p53 DNA-loaded nanoparticles demonstrated a sustained and significantly greater anti-proliferative effect than those transfected with naked wt-p53 DNA or wt-p53 DNA complexed with a commercially available transfecting agent (Lipofectamine). Additionally, cells transfected with wt-p53 DNA-loaded nanoparticles demonstrated sustained p53 mRNA levels compared with cells which were transfected with naked wt-p53 DNA or the wt-p53 DNA/Lipofectamine complex.[14]

Vascular endothelial growth factor (VEGF) has been investigated for the treatment of ischemic heart disease. PLGA nanoparticles encapsulating $VEGF_{165}$ pDNA were employed to explore the therapeutic potential of nanoparticles.[15] The nanoparticles, prepared by double emulsion and solvent evaporation technique, were observed to be of size ~100–300 nm, with DNA encapsulation efficiency of ~58%. Nanoparticles enabled significant enhancement of VEGF gene transfection into myocardial cells. TEM studies evidenced the presence of large numbers of nanoparticles in the myocardial cytoplasm and nucleus. Further, histological examination of ischemic rabbit myocardium transfected with nanoparticles exhibited a significant increase in the number of capillaries in contrast with the pDNA group.[15]

In an attempt to achieve high pDNA entrapment with full biological activity and minimal toxicity, PLGA nanoparticles were fabricated employing short sonication periods with low energy input, minimal washing steps, and without use of any additives.[16] The size of nanoparticles prepared using a modified water-in-oil-in-water (w/o/w) technique was found to be ~300 nm, with DNA entrapment efficiency of 80%. In view of results from the size distribution, DNA integrity, DNA entrapment efficiency, and release kinetics, it was concluded that the high MW system is preferable to the low MW system and that nanoparticle preparations involving shorter sonication times are preferable to preparations with long sonication times.[16] Further, the nanoparticles were non-toxic up to a concentration of 600 μg/ml in the cell lines tested, i.e. COS-7 and Cf2th. Confocal microscopy studies revealed that nanoparticles were localized in the endolysosomal compartment and later escaped into the cytosol. Transfection with nanoparticles significantly increased the protein expression of pGL3-Luc and was found to be 200-fold higher for COS-7 cells and 250-fold higher for Cf2th cells in comparison to control.[16]

The effect of pigment epithelial-derived factor (PEDF) gene loaded in PLGA nanoparticles was investigated on

Published by Woodhead Publishing Limited, 2013

mouse colon carcinoma cells (CT26) *in vitro* and *in vivo*.[17] Nanoparticles were found to measure ~220 nm, with 88% DNA encapsulation efficiency. Nanoparticles transfected to human umbilical vein endothelial cells (HUVEC) induced CT26 apoptosis and inhibited proliferation. Further, intratumoral administration of nanoparticles to mice models inhibited CT26 tumor growth by inducing CT26 apoptosis, decreasing microvessel density, and inhibiting angiogenesis.[17]

15.3 Chitosan-modified PLGA nanoparticles

Chitosan has been largely employed as main gene delivery vector in several studies. It can also be conjugated to other polymers to offer a flexible technology platform. The use of PLGA for DNA delivery application is limited by its negative surface charge and acidic degradation products. The incorporation of chitosan into the PLGA matrix could inhibit degradation of the matrix, thereby blocking the release of DNA from the polymer. Chitosan-coated nanoparticles are characterized by a strongly positive zeta potential, in contrast to the non-coated PLGA nanoparticles. Therefore, aggregation of these nanoparticles is prevented, as they are stabilized by strong electrostatic repulsion. To engineer cationic PLGA nanoparticles with defined size and shape that can efficiently bind DNA, an emulsion diffusion evaporation technique was employed with PVA-chitosan as stabilizer.[18] The cationic nanoparticles possessed a zeta potential of + 10 mV at pH 7.4 and size of ~185 nm. DNA retardation studies suggested the presence of sufficient charge on the nanoparticles to efficiently bind the negatively charged DNA electrostatically. Tahara et al. prepared pDNA-loaded PLGA nanospheres using the emulsion solvent diffusion in

Published by Woodhead Publishing Limited, 2013

which the surface was modified by binding to chitosan.[19] Chitosan-PLGA nanospheres were prepared by adding chitosan solution into the outer phase with PVA solution. The size of the nanospheres was found to be ~200–300 nm, with positive surface charge. Chitosan-modified or unmodified PLGA nanospheres did not show cytotoxicity with respect to the A549 cells, while DOTAP/pDNA complex exhibited ~20% cell death after 4 h of incubation. The luciferase gene expression mediated by nanospheres was initially weak, but gradually grew stronger, suggesting sustained release of pDNA from PLGA nanospheres. The improved transfection efficacy with chitosan-modified PLGA was thought to be due to endosomal escape of nanospheres, similar to that of chitosan/DNA nanoparticles.[19]

In another study, to improve pDNA loading efficiency and cellular uptake ability, chitosan was incorporated into the PLGA matrix.[20] PLGA-chitosan nanospheres were engineered by a spontaneous emulsion diffusion process and a particle size of around 60 nm was obtained under optimum formulation conditions. Due to the positive zeta potential of chitosan-modified PLGA nanospheres, they showed much higher loading efficiency than unmodified PLGA nanospheres. Transfection experiments in HepG2.2.15 cells showed a higher enhanced green fluorescent protein (EGFP) intensity after transfection with PLGA-chitosan-pDNA nanospheres (200 μg/2 μg) complexes in contrast to an equal amount of plain unmodified PLGA-pDNA nanospheres.[20]

15.4 Polyethylenimine-modified PLGA nanoparticles

PEI possesses strong electrostatic interaction with negatively charged DNA, and thereby confers protection on DNA

Published by Woodhead Publishing Limited, 2013

against enzymatic degradation, due to its high positive charge density. Hence, PEI has been widely investigated as a cationic polymer for the synthesis of cationic PLGA particles. In order to achieve efficient pulmonary gene expression, PLGA nanoparticles were formulated by adding the cationic polymer PEI.[21] PLGA-PEI nanoparticles prepared at different ratios and loaded with varying amounts of DNA showed particle sizes ranging between 207 and 231 nm with a narrow size distribution. All the nanoparticles exhibited high DNA loading efficiency (~99%) and were highly positive (zeta potential above 30 mV). DNA was observed in the endolysosomal compartment 6 h after DNA-loaded nanoparticles were transfected into the human airway submucosal epithelial cell line Calu-3. Further, the cytotoxicity of these nanoparticles was dependent on the PEI/DNA ratios, with highest cell viability at N/P ratios of 0.5 to 1.[21]

PLA/PEI and PLGA/PEI nanoparticles fabricated by a diafiltration method exhibited spherical and smooth surface characteristics.[22] The size of the nanoparticles was dependent on the amount of PEI, with reduction in size of PLA/PEI and PLGA/PEI nanoparticles from ~560 to 217 nm and ~580 to 185 nm, respectively, at wt/wt ratios of 0 to 0.6. Further, incorporation of PEI into PLA and PLGA nanoparticles rendered them stable for up to 12 days of storage at room temperature. The transfection efficiency in HEK 293 cells mediated by PLA and PLGA nanoparticles with PEI was ~50% of that with PEI alone and was significantly higher than that of naked DNA and nanoparticles without PEI.[22]

A combination of chemical (PLGA/PEI/DNA nanoparticles) and physical (ultrasonication) approaches was employed to improve the *in vivo* gene transfer efficiency via intravenous administration.[23] The size of PLGA/PEI/DNA nanoparticles was found to be ~200 nm, with positive zeta potential of

Published by Woodhead Publishing Limited, 2013

~13.4 mV. When intravenously administered in nude mice with DU145 human prostate tumors, the combination of PLGA/PEI/DNA nanoparticles with ultrasonication substantially enhanced *in vivo* transfection into tumor cells. The transfection efficacy was increased by up to eightfold in irradiated tumors compared with non-irradiated controls, while little to no cell death was produced by ultrasonication.[23]

To develop an effective vector for DNA vaccines, the gene expression and intracellular trafficking of pDNA complexed with PLGA/PEI nanospheres, in combination with an NF-κB analog as a nuclear localization signal (NLS) and electroporation, were evaluated in human monocyte-derived dendritic cells (hMoDCs).[24] PLGA/PEI nanospheres improved the cellular uptake of pDNA in both COS7 and hMoDCs cells. Further, the PLGA/PEI nanospheres significantly promoted transfection in COS-7 cells with high intranuclear transport of pDNA, but had almost no effect on transfection in hMoDCs cells with low intranuclear transport. However, PLGA/PEI nanospheres combined with NLS and electroporation (experimental permeation enhancer) significantly enhanced the transfection efficiency by improvement of intracellular uptake as well as intranuclear transport of pDNA in the hMoDCs.[24]

A PLGA nanoparticle formulation was optimized for gene delivery to a pulmonary epithelium model.[25] The surface properties of PLGA nanoparticles were varied by using different coating materials that adsorb to the particle surface during formation employing solvent diffusion methodology. Nanoparticles of size ~200 nm efficiently encapsulated plasmids encoding for luciferase (80–90%) and exhibited sustained release for 2 weeks. High levels of gene expression were observed at day 5 for positively charged PLGA nanoparticles that remained for at least 2 weeks, while with PEI/DNA complexes expression lasted only 5 days.

Published by Woodhead Publishing Limited, 2013

Furthermore, the PLGA nanoparticles were significantly less cytotoxic than PEI, suggesting the use of these vehicles for localized, sustained gene delivery to the pulmonary epithelium.[25]

To evaluate the influence of BPEI (25 kDa) incorporation into PLGA nanoparticles, two different PLGA nanoparticle-based gene delivery systems, namely, pDNA-encapsulated PLGA nanoparticles and pDNA adsorbed on the surface of PLGA-BPEI nanoparticles, were fabricated.[26] Nanoparticles of size 200 to 270 nm were observed, in which PLGA-BPEI (25 kDa) showed larger zeta potential and particle size than PLGA-BPEI (1.8 kDa). Further, the size of PLGA-BPEI-(25 kDa)/pDNA complexes gradually increased from 135 nm to 463 nm during an incubation time of 120 min. Transfection studies in HEK293 cells exhibited a six to sevenfold increase with PLGA-BPEI NPs as compared with those without PEI at day 2, but with significantly higher cytotoxicity.[26] Intracellular tracking studies revealed PLGA-BPEI nanoparticle localization in the nucleus, while nanoparticles without PEI were seen in the cytoplasm. Additionally, pDNA release from pDNA-encapsulated PLGA nanoparticles was relatively faster than from PLGA-BPEI nanoparticles, but lower gene transfection was observed because of the lack of active transport of the released pDNA to the nucleus.[26]

The differentiation capability of human MSC cells, using the SOX trio genes as targets, was evaluated after modification with biodegradable PLGA nanoparticles.[27] The sizes of the PLGA nanoparticles were ~60, 72, and 92 nm, with SOX5, 6, and 9 genes fused to green fluorescence protein (GFP), yellow fluorescence protein (YFP), or red fluorescence protein (RFP), respectively. *In vitro* transfection studies done in MSC cells using SOX trio complexed with PEI-modified PLGA nanoparticles led to a dramatic increase in

Published by Woodhead Publishing Limited, 2013

chondrogenesis. For the PEI/GFP and PEI/SOX5, 6, and 9 genes complexed with PLGA nanoparticles, the expression of GFP as a reporter gene and SOX9 genes with PLGA nanoparticles showed 80% and 83% of gene transfection ratios into hMSCs 2 days after transfection, respectively.[27]

Shau et al. compared a one-step nanoparticle preparation process of PLGA pre-modified with PEI with those prepared in a two-step process (preformed PLGA nanoparticles subsequently coated with PEI).[28] The nanoparticles were prepared by emulsification of PLGA/ethyl acetate in an aqueous solution of PVA and PEI. Physico-chemical studies showed that the PLGA/PEI nanoparticles were spherical and non-porous with a size of ~200 nm and positive zeta potential. Also, the zeta potential of the nanoparticles from the one-step procedure was markedly higher than that from the two-step process, which was ascribed to the conjugation of PEI to PLGA via aminolysis. The PLGA/PEI nanoparticles complexed with DNA possessed significantly lower cytotoxicity and a higher transfection activity than PEI polyplexes.[28]

15.5 Other modifications to PLGA nanoparticles

The surface of PLGA nanoparticles has been manipulated with various ligands to improve upon the gene delivery efficiency. The modifications consist of either imparting cationic charge or improving the targetability of PLGA nanoparticles. To develop a liver-targeted gene delivery system, asialofetuin (AF), a glycoprotein with triantennary galactose terminal sugar chains, was complexed with PLGA/DOTAP nanoparticles prepared by the solvent evaporation method, with DNA either encapsulated or adsorbed onto

Published by Woodhead Publishing Limited, 2013

nanoparticles.[29] The size of the nanoparticles ranged from ~425 to 671 nm, with unimodal size distribution (polydispersity index <0.25) and positive zeta potential along with DNA loading efficiency of up to 99%. Targeted nanoparticles loaded with genes encoding for luciferase and interleukin-12 (IL-12) resulted in increased transfection efficiencies in HepG2 cells as compared with free DNA and with plain (non-targeted) systems, even in the presence of 60% FBS. Furthermore, *in vivo* studies suggested that the presence of the targeting ligand increases transfection efficiency by 20-fold in the liver as compared with the lung.[29]

Novel bioadhesive PLGA nanoparticles were designed for efficient gene delivery to lung cancer cells using a bioadhesive agent and stabilizer, Carbopol 940 and Pluronic F68, with Pluronic F127-stabilized PLGA nanoparticles as control.[30] The size of the nanoparticles decreased from 198 to 126 nm when the concentration of Carbopol 940 was increased from 0.005 to 0.02% (w/v), but increased with further increase in the concentration of Carbopol from 0.02 to 0.1%. All the obtained nanoparticles showed negative surface charge, similar spherical morphology, a relatively narrow particle size distribution, and lower cytotoxicity to A549 cells in comparison with Lipofectamine 2000. The nanoparticles showed higher transfection efficiency in A549 cells as compared with Pluronic-stabilized nanoparticles or naked DNA, similar to that of Lipofectamine 2000.[30]

Fay et al. developed PLGA nanoparticles to encapsulate DNA, using a combination of salting out and emulsion–evaporation processes.[31] Highly monodispersed nanoparticles of size ~240 nm were produced, which entrapped DNA in both supercoiled and open circular form. Further, to promote endosomal escape of the nanoparticles, the cationic charged surfactants cetyl trimethylammonium bromide (CTAB) and dimethyldidodecylammonium bromide (DMAB)

Published by Woodhead Publishing Limited, 2013

were introduced, which led to an increase in zeta potential. Introduction of positive charge reduced the cell viability, while, at the same positive zeta potential, the DMAB-coated nanoparticles induced significantly less cytotoxicity than those with CTAB. Furthermore, intracellular tracking studies revealed enhanced endolysosomal escape of DMAB-coated nanoparticles, followed by localization in the cytosol. Also, DMAB-coated PLGA nanoparticles loaded with a GFP reporter plasmid exhibited significant improvements in transfection efficiencies (up to 20%) in comparison to non-modified particles, highlighting their functional usefulness.[31]

The surface of PLGA nanospheres was modified with polysorbate 80 (P80) in order to improve the cellular uptake and transfection efficiency.[32] The size of PLGA nanospheres ranged from 250 to 300 nm, depending on the type of surface modifier used, i.e. chitosan or P80. The cellular uptake of P80-PLGA nanospheres in A549 cells was much higher than that of unmodified and chitosan-modified PLGA nanospheres. Also, the luciferase gene expression level obtained with P80-PLGA nanospheres was significantly higher than that of unmodified and chitosan-modified PLGA nanospheres. Further, the uptake of unmodified and chitosan-modified nanospheres was mediated, predominantly, by clathrin-mediated endocytosis, while the pathway for P80-PLGA nanospheres could not be determined.[32]

In order to enhance incorporation efficiency and improve release kinetics of pDNA from PLGA nanoparticles, a facile method for the fabrication of calcium phosphate-embedded PLGA nanoparticles was developed.[33] The nanoparticles were observed to be spherical with size of ~200 nm and an entrapment efficiency of ~95.7%. The surface area measurement studies revealed that the nanoparticles were mesoporous in nature, with specific surface area of 57.5 m^2/g and an average pore size of 96.5 Å. Further, the nanoparticles

Published by Woodhead Publishing Limited, 2013

exhibited 22.4% transfection in HEK 293 cells, which was considerably higher than that of both the pDNA-loaded PLGA nanoparticles and the calcium phosphate-pDNA-embedded PLGA microparticles.[33]

Poly(beta-amino esters) (PBAEs) have the ability to form hybrid particles with other polymers, which allows the production of solid, stable, and storable particles. Hybrid PBAE/PLGA nanoparticles were synthesized and optimized for uptake by and transfection of cystic fibrosis-affected bronchiolar epithelial cells (CFBE41o- cells).[34] Efficient cellular internalization and transfection (COS-7 and CFBE41o-) were observed with the nanoparticles containing PBAE as compared with those without PBAE, and they also exhibited higher cytotoxicity. The toxicity of the nanoparticles was circumvented by coating the surface with the CPPs mTAT, bPrPp and MPG via a PEGylated phospholipid linker (DSPE-PEG2000). Further, the nanoparticles modified with bPrPp and MPG coating resulted in 3–4.5-fold more pDNA loading than unmodified particles and approximately an order of magnitude improvement on transfection efficiency in CFBE41o- cells.[34]

15.6 Conclusions

PLGA nanoparticles can improve the therapeutic efficacy of gene delivery due to sustained and controlled release of the encapsulated DNA over a period of days to several weeks compared with natural polymers. Further, PLGA has been approved by the US FDA and the European Medicine Agency in drug delivery systems for parenteral administration. However, thorough understanding of pharmacokinetics, biodistribution, toxicity, and efficacy of gene delivery is still required before widespread use of PLGA nanoparticles in

Published by Woodhead Publishing Limited, 2013

clinical trials. Nevertheless, gene delivery employing PLGA or PLGA-based nanoparticles is a potential area for biomedical research.

15.7 References

1. Merdan, T., Kopecek, J. and Kissel, T. (2002) Prospects for cationic polymers in gene and oligonucleotide therapy against cancer. *Adv Drug Deliv Rev*, **54**, 715–58.
2. Kommareddy, S. and Amiji, M. (2005) Preparation and evaluation of thiol-modified gelatin nanoparticles for intracellular DNA delivery in response to glutathione. *Bioconjug Chem*, **16**, 1423–32.
3. Lemieux, P., Guerin, N., Paradis, G., Proulx, R., Chistyakova, L., et al. (2000) A combination of poloxamers increases gene expression of plasmid DNA in skeletal muscle. *Gene Ther*, **7**, 986–91.
4. Morille, M., Passirani, C., Vonarbourg, A., Clavreul, A. and Benoit, J.P. (2008) Progress in developing cationic vectors for non-viral systemic gene therapy against cancer. *Biomaterials*, **29**, 3477–96.
5. Bhavsar, M.D. and Amiji, M.M. (2007) Polymeric nano- and microparticle technologies for oral gene delivery. *Expert Opin Drug Deliv*, **4**, 197–213.
6. Panyam, J. and Labhasetwar, V. (2003) Biodegradable nanoparticles for drug and gene delivery to cells and tissue. *Adv Drug Deliv Rev*, **55**, 329–47.
7. Panyam, J., Zhou, W.Z., Prabha, S., Sahoo, S.K. and Labhasetwar, V. (2002) Rapid endo-lysosomal escape of poly(DL-lactide-co-glycolide) nanoparticles: implications for drug and gene delivery. *FASEB J*, **16**, 1217–26.

Published by Woodhead Publishing Limited, 2013

8. Panyam, J. and Labhasetwar, V. (2004) Sustained cytoplasmic delivery of drugs with intracellular receptors using biodegradable nanoparticles. *Mol Pharm*, **1**, 77–84.
9. Cohen, H., Levy, R.J., Gao, J., Fishbein, I., Kousaev, V., et al. (2000) Sustained delivery and expression of DNA encapsulated in polymeric nanoparticles. *Gene Ther*, **7**, 1896–905.
10. Cohen-Sacks, H., Najajreh, Y., Tchaikovski, V., Gao, G., Elazer, V., et al. (2002) Novel PDGFbetaR antisense encapsulated in polymeric nanospheres for the treatment of restenosis. *Gene Ther*, **9**, 1607–16.
11. Prabha, S., Zhou, W.Z., Panyam, J. and Labhasetwar, V. (2002) Size-dependency of nanoparticle-mediated gene transfection: studies with fractionated nanoparticles. *Int J Pharm*, **244**, 105–15.
12. Muller, R.H. (ed.) (1991) *Colloidal carriers for controlled drug delivery and targeting*. CRC Press, Boca Raton.
13. Panyam, J., Sahoo, S.K., Prabha, S., Bargar, T. and Labhasetwar, V. (2003) Fluorescence and electron microscopy probes for cellular and tissue uptake of poly(D,L-lactide-co-glycolide) nanoparticles. *Int J Pharm*, **262**, 1–11.
14. Prabha, S. and Labhasetwar, V. (2004) Nanoparticle-mediated wild-type p53 gene delivery results in sustained antiproliferative activity in breast cancer cells. *Mol Pharm*, **1**, 211–19.
15. Yi, F., Wu, H. and Jia, G.L. (2006) Formulation and characterization of poly (D,L-lactide-co-glycolide) nanoparticle containing vascular endothelial growth factor for gene delivery. *J Clin Pharm Ther*, **31**, 43–8.
16. Gvili, K., Benny, O., Danino, D. and Machluf, M. (2007) Poly(D,L-lactide-co-glycolide acid) nanoparticles for

DNA delivery: waiving preparation complexity and increasing efficiency. *Biopolymers*, **85**, 379–91.

17. Cui, F.Y., Song, X.R., Li, Z.Y., Li, S.Z., Mu, B., et al. (2010) The pigment epithelial-derived factor gene loaded in PLGA nanoparticles for therapy of colon carcinoma. *Oncol Rep*, **24**, 661–8.
18. Ravi Kumar, M.N., Bakowsky, U. and Lehr, C.M. (2004) Preparation and characterization of cationic PLGA nanospheres as DNA carriers. *Biomaterials*, **25**, 1771–7.
19. Tahara, K., Sakai, T., Yamamoto, H., Takeuchi, H., Hirashima, N., et al. (2011) Improvements in transfection efficiency with chitosan modified poly(D, L-lactide-co-glycolide) nanospheres prepared by the emulsion solvent diffusion method, for gene delivery. *Chem Pharm Bull (Tokyo)*, **59**, 298–301.
20. Zeng, P., Xu, Y., Zeng, C., Ren, H. and Peng, M. (2011) Chitosan-modified poly(d,l-lactide-co-glycolide) nanospheres for plasmid DNA delivery and HBV gene-silencing. *Int J Pharm*, **415**, 259–66.
21. Bivas-Benita, M., Romeijn, S., Junginger, H.E. and Borchard, G. (2004) PLGA-PEI nanoparticles for gene delivery to pulmonary epithelium. *Eur J Pharm Biopharm*, **58**, 1–6.
22. Kim, I.S., Lee, S.K., Park, Y.M., Lee, Y.B., Shin, S.C., et al. (2005) Physicochemical characterization of poly(L-lactic acid) and poly(D,L-lactide-co-glycolide) nanoparticles with polyethylenimine as gene delivery carrier. *Int J Pharm*, **298**, 255–62.
23. Chumakova, O.V., Liopo, A.V., Andreev, V.G., Cicenaite, I., Evers, B.M., et al. (2008) Composition of PLGA and PEI/DNA nanoparticles improves ultrasound-mediated gene delivery in solid tumors in vivo. *Cancer Lett*, **261**, 215–25.

Published by Woodhead Publishing Limited, 2013

24. Kanazawa, T., Takashima, Y., Murakoshi, M., Nakai, Y. and Okada, H. (2009) Enhancement of gene transfection into human dendritic cells using cationic PLGA nanospheres with a synthesized nuclear localization signal. *Int J Pharm*, **379**, 187–95.
25. Baoum, A., Dhillon, N., Buch, S. and Berkland, C. (2010) Cationic surface modification of PLG nanoparticles offers sustained gene delivery to pulmonary epithelial cells. *J Pharm Sci*, **99**, 2413–22.
26. Son, S. and Kim, W.J. (2010) Biodegradable nanoparticles modified by branched polyethylenimine for plasmid DNA delivery. *Biomaterials*, **31**, 133–43.
27. Park, J.S., Yang, H.N., Woo, D.G., Jeon, S.Y., Do, H.J., et al. (2011) Chondrogenesis of human mesenchymal stem cells mediated by the combination of SOX trio SOX5, 6, and 9 genes complexed with PEI-modified PLGA nanoparticles. *Biomaterials*, **32**, 3679–88.
28. Shau, M.D., Shih, M.F., Lin, C.C., Chuang, I.C., Hung, W.C., et al. (2012) A one-step process in preparation of cationic nanoparticles with poly(lactide-co-glycolide)-containing polyethylenimine gives efficient gene delivery. *Eur J Pharm Sci*, **46**, 522–9.
29. Díez, S., Miguéliz, I. and Tros de Ilarduya, C. (2009) Targeted cationic poly(D,L-lactic-co-glycolic acid) nanoparticles for gene delivery to cultured cells. *Cell Mol Biol Lett*, **14**, 347–62.
30. Zou, W., Liu, C., Chen, Z. and Zhang, N. (2009) Studies on bioadhesive PLGA nanoparticles: A promising gene delivery system for efficient gene therapy to lung cancer. *Int J Pharm*, **370**, 187–95.
31. Fay, F., Quinn, D.J., Gilmore, B.F., McCarron, P.A. and Scott, C.J. (2010) Gene delivery using dimethyldidodecylammonium bromide-coated PLGA nanoparticles. *Biomaterials*, **31**, 4214–22.

Published by Woodhead Publishing Limited, 2013

32. Tahara, K., Yamamoto, H. and Kawashima, Y. (2010) Cellular uptake mechanisms and intracellular distributions of polysorbate 80-modified poly (D,L-lactide-co-glycolide) nanospheres for gene delivery. *Eur J Pharm Biopharm*, **75**, 218–24.
33. Tang, J., Chen, J.Y., Liu, J., Luo, M., Wang, Y.J., et al. (2012) Calcium phosphate embedded PLGA nanoparticles: a promising gene delivery vector with high gene loading and transfection efficiency. *Int J Pharm*, **431**, 210–21.
34. Fields, R.J., Cheng, C.J., Quijano, E., Weller, C., Kristofik, N., et al. (2012) Surface modified poly(beta amino ester)-containing nanoparticles for plasmid DNA delivery. *J Control Release*, **164**, 41–8.

Published by Woodhead Publishing Limited, 2013

16

Metallic and inorganic nanoparticles

DOI: 10.1533/9781908818645.331

Abstract: Noble metal based nanoparticles exhibit exciting optical, catalytic, and magnetic properties. Gold nanoparticles have emerged as one of the delivery systems with most potential for efficient transport and release of pharmaceuticals into diverse cell types. The gold core is essentially inert, non-toxic, and biocompatible, which makes it an ideal material for vector design. Gold nanoparticles can be coupled to DNA either covalently or by non-covalent interactions. Silica with well defined physico-chemical properties such as chemical structures and surface properties are known to be biocompatible. Furthermore, the silanol surface chemistry of silica nanoparticles provides ease of modification and functionalization for enhanced nucleic acids delivery.

Key words: gold nanoparticles, thiol, covalent conjugation, non-covalent conjugation, mesoporous silica nanoparticles, silanol groups.

Published by Woodhead Publishing Limited, 2013

16.1 Introduction

Targeted and programmed delivery of therapeutic molecules to specific physiological targets is a key challenge for molecular and macromolecular therapeutics. During the past few decades several nanocarriers, including liposomes, polymeric nanoparticles, polymer micelles, dendrimers, metal nanoparticles, and silica nanoparticles, have been used as promising gene delivery vectors. Noble metal-based nanoparticles exhibit fascinating optical, catalytic, and magnetic properties that are distinct from their bulk counterparts. Gold nanoparticles have emerged as one of the delivery systems with most potential for efficient transport and release of pharmaceuticals into diverse cell types. Silicas with well-defined physico-chemical properties, such as chemical structures and surface properties, are known to be biocompatible. Among inorganic nanoparticles, silica nanoparticles are often the material of choice for biological applications.

16.2 Gold nanoparticles

AuNPs exhibit several desirable physico-chemical properties that make them preferred candidates for gene delivery applications. The gold core is essentially inert, non-toxic, and biocompatible, which makes it an ideal material for vector design.[1] AuNPs can be easily synthesized with controlled dispersity in a wide range of sizes (1–150 nm).[2] Size and size distribution play a key role in nanoparticle-mediated gene delivery. Also, size and properties of AuNPs can be manipulated to match the size of biomolecules such as proteins and DNA, facilitating their integration into biological systems. Moreover, the high surface area-to-

Published by Woodhead Publishing Limited, 2013

volume ratio of nanoparticles provides high loading of functionalities, including targeting and therapeutic moieties.[3] For instance, AuNPs 2 nm in diameter can accommodate ~100 ligands via covalent conjugation.[4] Finally, the versatility of modulating and multivalent surface structures of AuNPs allows incorporation of diverse therapeutic drugs or biomacromolecules by covalent or non-covalent conjugation on the surface of nanoparticles.[5,6] A variety of functional monolayers can be created on the surface of AuNPs to facilitate payload release strategies using internal or external stimuli such as glutathione, pH, heat, and light. AuNPs are not restricted to being vectors of only small molecules, but can efficiently deliver large biomolecules. Size and functionality modulation capabilities make them a useful system for efficient recognition and delivery of biomolecules.

16.2.1 Synthesis of gold nanoparticles

Sincere efforts have been dedicated towards fabrication of AuNPs with controlled size and monodispersity. AuNPs with varying sizes have been prepared by the reduction of gold salts in the presence of appropriate stabilizing agents that prevent particle agglomeration. Some of the commonly used synthetic methods for AuNPs are summarized in Table 16.1. Several studies report synthesis of AuNP-based delivery systems bearing functional moieties, which are usually conjugated with thiol-linkers, in their monolayers. Schiffrin et al. developed a one-pot synthesis approach for rapid and scalable preparation of a wide variety of monolayer-protected clusters (MPCs) (Figure 16.1).[7] This strategy comprises reduction of $AuCl_4^-$ salts with $NaBH_4$ in the presence of the desired thiol capping ligand or ligands. Further, the size of the core particles can be varied from 1.5 nm to ~6 nm by tweaking the thiol–gold stoichiometry.

Published by Woodhead Publishing Limited, 2013

Table 16.1 Commonly used synthetic methods for preparation of AuNPs

Core size	Synthetic methods	Capping agents	References
1–2 nm	Reduction of $AuCl(PPh_3)$ with diborane or sodium borohydride	Phosphine	Schmid[8]
1.5–5 nm	Biphasic reduction of $HAuCl_4$ by sodium borohydride in the presence of thiol capping agents	Alkanethiol	Brust et al.,[7] Templeton et al.[9]
10–150 nm	Reduction of $HAuCl_4$ with sodium citrate in water	Citrate	Grabar et al.,[10] Frens,[11] Turkevich et al.[12]

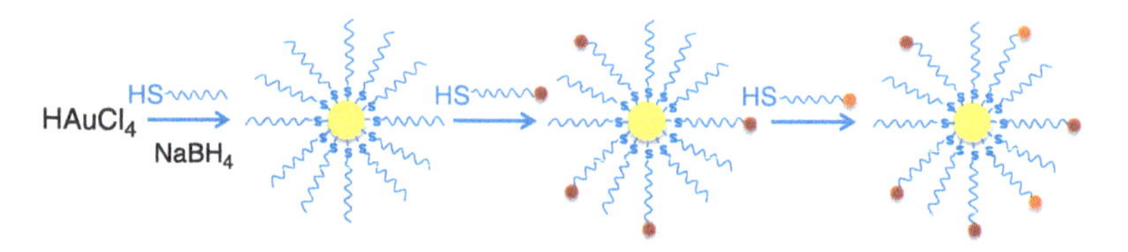

Figure 16.1 Formation of MPCs using the Schiffrin reaction and mixed MPCs using the Murray's place-exchange reaction

16.2.2 Gold nanoparticles for gene delivery via non-covalent conjugation of DNA

In one of the initial studies, mixed monolayer protected gold clusters (MMPCs) functionalized with tetraalkylammonium ligands were fabricated with a size of 6–10 nm, which can interact with the DNA backbone via charge complementarity.[13] The stoichiometry of the DNA–AuNP complexation process suggested that four nanoparticles could bind to each DNA strand, two on each "side" of the DNA strand. Further, the interactions of DNA-AuNPs completely inhibited

Published by Woodhead Publishing Limited, 2013

transcription by T7 RNA polymerase *in vitro*. Furthermore, these nanoparticles efficiently transfected HEK 293 cells with pDNA coding for β-galactosidase gene.[14] The transfection efficiency was dependent on the ratio of DNA to nanoparticle during the incubation period, the number of charged substituents in the monolayer core, and the hydrophobic packing surrounding these amines. The nanoparticle/DNA complexes prepared at w/w ratios of 30 were observed to be most effective in promoting transfection in the presence of 10% serum and 100 μM chloroquine.[14] In another study, cationic trimethylammonium-functionalized AuNPs also complexed with DNA through electrostatic interactions, resulting in complete inhibition of DNA transcription of T7 RNA polymerase.[15] The intracellular concentrations of glutathione promoted release of DNA from the nanoparticle, resulting in efficient transcription. Additionally, the nanoparticle-bound DNA was efficiently protected from DNAse I digestion and in a physical sonication assay.[16] However, the AuNP-bound DNA was found to show enhanced cleavage upon exposure to chemically induced radicals.

Positively charged AuNPs, of size 1.4 nm with approximately six primary amine groups, complexed with pDNA encoding for murine interleukin-2 (pVAXmIL-2), were employed to enhance both gene delivery and transfection efficiency.[17] The cellular delivery to C2C12 cells was enhanced 6.3-fold with AuNP/DNA complexes as compared with PEI(25 kDa)/DNA complexes. Further, the AuNP/DNA exhibited 3.2- and 2.1-fold higher murine IL-2 protein expression than that observed with PEI(25 kDa)/DNA and PEI(2 kDa)/DNA complexes, respectively. Furthermore, following intramuscular administration, AuNP/DNA complexes showed expression levels more than four orders of magnitude higher as compared with naked DNA.[17]

Thomas et al. covalently conjugated BPEI (MW 2 kDa) to AuNPs to prepare *in vitro* transfection vectors into monkey kidney (COS-7) cells in the presence of serum.[18] The transfection efficiencies were found to be dependent on the PEI/gold molar ratio in the conjugates; at optimal ratio the conjugates were ~12 times more potent than the unmodified polycation. Further, the transfection efficiency was increased twofold by addition of amphiphilic N-dodecyl-PEI during complex formation with DNA. Also, unmodified PEI transfected just 4% of the cells, while PEI-AuNPs transfected 25%, and the PEI-AuNPsI/dodecyl-PEI ternary complex transfected 50% of the cells. Furthermore, the intracellular trafficking of the nanoparticle/DNA complexes led to their presence in the nucleus <1 h after transfection.[18] Later, Wang et al. prepared efficient DNA binders by conjugation of β-CD to the surface of oligo(ethylenediamino)-modified AuNPs.[19] The nanoparticle/DNA complexes effectively transfected human breast cancer cells (MCF-7) and the level of gene expression was dependent on the w/w ratio between nanoparticles and DNA.

To improve the efficiency of reverse transfection in MSC cells, AuNPs (~20 nm) were coupled with commercially available Jet-PEI.[20] The AuNPs-PEI/DNA conjugates showed more than a 2.5-fold increase in transfection efficiency as compared with control PEI/DNA complexes without AuNPs. PEI (800 Da) conjugated with AuNPs completely retarded pDNA at N/P ratios above 4 in electrophoresis on agarose gel, and also provided effective protection of DNA against DNase.[21] AuNPs-PEI with higher PEI grafting density resulted in more compact and smaller complexes with plasmid DNA, as compared with those with lower grafting density. In the absence of serum, AuNPs-PEI transfected pGL-3 to COS-7 cells three to fourfold more efficiently than unmodified PEI800. Further, the transfection efficiency of

Published by Woodhead Publishing Limited, 2013

AuNPs-PEI remained unaltered with addition of serum in media, while that of PEI (25 kDa) decreased markedly. The transfection efficiency of AuNPs-PEI was ~60-fold that of PEI (25 kDa) in 10% serum medium and exhibited only mild toxicity.[21]

In order to gain insight into the influence of particle size on cell transfection, two different sizes of Au-PEI nanoparticles of ~6 and 70 nm were complexed with pDNA carrying reporter or suicide genes to prepare Au-PEI/DNA complexes, and transfection efficacy was investigated in human osteosarcoma Saos-2 cells. Results revealed efficient transfection with complexes derived from ~6 nm Au-PEI nanoparticles as compared with ~70 nm Au-PEI nanoparticles. Further investigation showed the presence of large aggregates of ~70 nm nanoparticles in endocytic vesicles of cells, while smaller nanoparticles showed lower agglomeration and effective endosomal escape.[22] Colocalization studies have given evidence of presence of DNA inside the cell nucleus after 2 h of transfection with ~6 nm Au-PEI nanoparticles, and it remained present in the sub-cellular compartment for up to 3 days.[22]

Zhou et al. covalently coupled low MW chitosan (MW 6 kDa) to AuNPs by reduction of $HAuCl_4$–chitosan mixtures with sodium borohydride, and investigated *in vitro* and *in vivo* potency of the resulting chitosan-AuNP conjugates as pDNA delivery vectors.[23] The chitosan-AuNPs/DNA complexes appeared by AFM as multi-domain structures of irregular shape with size of ~250–450 nm. *In vitro* transfection experiments performed in RAW264.7 cells exhibited relatively high transfection efficiencies for chitosan-AuNPs/DNA complexes as compared with chitosan only, which could be attributed to the stability of the complexes in 10% serum. Further, intramuscular immunization in BALB/c mice with the chitosan-AuNPs/

Published by Woodhead Publishing Limited, 2013

DNA complexes induced an enhanced serum antibody response, ten times more potent than with naked DNA vaccine. Furthermore, chitosan-AuNPs/DNA complexes induced potent cytotoxic T lymphocyte responses at a low dose, in contrast to naked DNA.[23]

Han et al. fabricated cationic AuNPs with a photoactive *o*-nitrobenzyl ester linkage and a quaternary ammonium salt as endgroup, which allowed temporal and spatial release of DNA by light.[24] After complexation of the nanoparticles with DNA, near-UV irradiation (>350 nm) was used to cleave the nitrobenzyl linkage, resulting in formation of a negatively charged carboxylate group and release of positively charged alkyl amine. *In vitro* studies using a T7 RNA polymerase assay demonstrated that DNA transcription was restored upon UV irradiation. Further, uptake studies in cell culture showed efficient internalization of nanoparticle/DNA complexes. Release of FITC-labeled DNA upon irradiation was established by observation of bright fluorescence inside the cells.[24]

16.2.3 Gold nanoparticles for gene delivery via covalent linkage of DNA

DNA can be easily modified with thiols (-SH) for covalent conjugation onto the surface of AuNPs. Rosi et al. conjugated ODNs to AuNPs by modification of ODNs via thiol linkers. The ODN-nanoparticles with high negative charge were significantly internalized by a mouse endothelial cell line (C166 cells).[25] These complexes were stable against enzymatic digestion and showed up to 99% internalization in C166 cells. Antisense efficacy of ODN-nanoparticle complexes was demonstrated using EGFP-expressing C166, resulting in reduction of fluorescence intensity in cells.[25] Further, the mechanistic studies of cellular uptake of ODN-nanoparticles

Published by Woodhead Publishing Limited, 2013

exhibited that the endocytosis process was initiated by adsorption of several serum proteins onto the nanoparticle surface.[26]

16.3 Mesoporous silica nanoparticles

Silica-based nanoparticles have been investigated as nucleic acid delivery vectors due to their inherent biocompatibility and low toxicity. Moreover, the silanol surface chemistry of silica nanoparticles provides ease of modification and functionalization for enhanced nucleic acid delivery. Typically, MSNs possess an average particle diameter ~100 nm, surface area ~900 m^2/g, pore sizes ~2 nm, pore volumes ~0.9 cm^3/g, abundant surface silanol groups (~30 mol%), and a hexagonal porous channel structure. Properties of MSN that have rendered them suitable for drug and gene delivery application include:

- **Tunable particle size:** the particle size of MSN can be modulated between 50 and 300 nm, which facilitates cellular endocytosis without any significant cytotoxicity.
- **High stability:** MSN is highly resistant to heat, pH, mechanical stress, and hydrolysis-induced degradations as compared with other polymer-based vectors.
- **Uniform and tunable pore size:** MSNs have a very narrow pore size distribution, and pore diameter can be tuned between 2 and 6 nm.
- **High surface area and large pore volume:** MSNs possess high surface area (>900 m^2/g) and pore volume (>0.9 cm^3/g), which allows high loadings of drug molecules.
- **Two functional surfaces:** MSNs have an internal surface (i.e. cylindrical pores) and an external surface (i.e. exterior

particle surface). This characteristic allows the selective functionalization of the internal and/or external surfaces of MSN with different moieties.

16.4 MSN for gene delivery

Several advantageous properties of MSNs (including a narrow pore size distribution that can efficiently entrap DNA and allow release at target sites, and ease of surface functionalization to optimize plasmid adsorption and release characteristics) have made them potential candidates for gene delivery. Various studies have reported adsorption of plasmids into and onto MSNs, and evaluated them under *in vitro* conditions. However, the loading of DNA into the mesopores of MSNs is simple, and several studies report only adsorption of DNA to the outer surface of the cationically modified MSNs.

In one of the initial studies, second-generation (G2) PAMAM dendrimers were covalently attached to the surface of a MCM-41-type MSN of size 250 nm with average pore diameter of 2.7 nm.[27] MSNs complexed with a pDNA (pEGFP-C1) coding for an enhanced green fluorescence protein efficiently condensed DNA and protected against DNAse degradation. Transfection studies in HeLa cells revealed ~35% transfected cells with MSN, while only ~15, 10 and 16% transfection was seen with PolyFect, SuperFect, and Metafectene, respectively. Further, internalization studies exhibited endocytosis of a large number of MSN/DNA complexes followed by localization in sub-cellular organelles such as mitochondria and Golgi.[27]

Suwalski et al. reported the ability of amino- and carboxyl-modified MSN to deliver gene *in vivo* in rat Achilles tendons.[28] Luciferase gene expression was observed for at least 2 weeks

Published by Woodhead Publishing Limited, 2013

in tendons injected with MSN/pDNA encoding the luciferase reporter gene, while it was undetectable with naked DNA. Interestingly, after weekly administration of MSN/pDNA formulation for 1.5 months, rats showed no signs of inflammation or necrosis in tendon, kidney, heart, and liver. Further, the tendons treated with MSN/platelet-derived growth factor (PDGF) gene healed significantly faster than untreated tendons and those treated with pPDGF alone.[28]

To evaluate the capability of porous MSN with large pores that can accommodate large pDNAs by providing more inner space and allow better protection against nucleases, MSNs of size 250 nm with almost ten times the usual pore size (mean pore size 23 nm) were synthesized.[29] TEM and confocal microscopy studies revealed endocytosis of a large number of MSNs by HeLa cells on exposure to 80 μg/ml of MSN. Further, the size of the MSNs increased after incubation with DNA from 905 to 1223 nm and from 543 to 1071 nm for small-pore MSN and large-pore MSN, respectively. Also, the DNA loading capacity of large-pore MSNs was always higher than that of lower-pore size MSNs. Cell viability assays revealed that MSNs did not induce noticeable cell death in HeLa cells even at high concentration (640 μg/ml). Although the transfection efficiency of Lipofectamine 2000 was 3.5-fold higher than that of MSNs, it was also associated with higher cell death.[29]

Gao et al. adsorbed up to about 150 μg/g MSN of firefly luciferase plasmid DNA (5256 base-pairs) in PBS buffer.[30] The surface amino groups allowed an attractive interaction with the negatively charged DNA. The adsorbed plasmid was protected against enzymatic degradation, although the exact location of the plasmid, inside the mesopores and on the outer particle surface, remained unclear. Kim et al. reported complexation of amine-functionalized MSNs with bone morphogenetic protein-2 (BMP2) pDNA to study their transfection efficiency in MSC cells.[31] Significant intracellular uptake of the MSN/DNA complex occurred, with a transfection efficiency of ~68%. Furthermore, over 66% of the transfected cells produced BMP2 protein. The osteogenic differentiation of the transfected MSC cells was demonstrated by the expression of bone-related genes and proteins, including bone sialoprotein, osteopontin, and osteocalcin.[31]

Yu et al. investigated the acute toxicity of silica nanoparticles by systematically varying geometry, porosity, and surface characteristics by intravenous administration in immune-competent mice.[32] The toxicity was observed to be mainly influenced by nanoparticle porosity and surface characteristics, such that maximum tolerated dose (MTD) increased in the following order: mesoporous SiO_2 (aspect ratio 1, 2, 8) at 30–65 mg/kg < amine-modified mesoporous SiO_2 (aspect ratio 1, 2, 8) at 100–150 mg/kg < unmodified or amine-modified nonporous SiO_2 at 450 mg/kg. The adverse reactions above MTDs were primarily caused by the mechanical obstruction of the vasculature by SiO_2, which led to congestion in multiple vital organs and subsequent organ failure.[32]

16.5 Polycation-modified MSN for gene delivery

For successful application of large-pore MSNs in gene delivery, surface modification of the silica is required in order to generate sufficient binding affinity for the negatively charged DNA. Various techniques have been developed for incorporation of positively charged functional groups onto silica nanoparticles, either through non-covalent interactions or by covalent bonding. Li et al. used electrostatic interactions to complex PLL with silica nanoparticles and employed this

Published by Woodhead Publishing Limited, 2013

system for pDNA delivery.[33] The complexes efficiently condensed DNA, provided protection against enzymatic degradation, and consequently delivered AS-ODNs selectively into cells.[34] However, simple adsorption of PLL could not be used to uniformly modify silica particles larger than 60 nm, possibly due to the unfavorable surface energy of large silica particles.[33] Lunn et al. used surface-initiated polymerization of *N*-carboxyanhydrides (NCAs) to graft peptide (PLL and poly-l-alanine) onto the surface of amine-functionalized mesoporous silica materials (MCM-41, SBA-15, and KIT-6).[35] However, this functionalization procedure resulted in significant reduction of the porosity of the silica particles and pore blockage at high peptide loading. Consequently, the loading capacity of the particles was dramatically decreased.

Hartono et al. synthesized large-pore MSNs of size 100–200 nm with cage-like pores organized in a cubic mesostructure; the size of the cavities was ~28 nm, with an entrance size of 13.4 nm.[36] PLL-functionalized nanoparticles exhibited a strong ability to deliver oligo DNA-Cy3 (a model for siRNA) to HeLa cells. Furthermore, PLL-functionalized nanoparticles were proven to be superior as gene carriers compared with amino-functionalized nanoparticles and the native nanoparticles. Moreover, the PLL-modified silica nanoparticles also exhibited a high biocompatibility, with low cytotoxicity observed up to 100 μg/ml.[36] Bhattarai et al. modified the surface of MSNs with PEG and poly(2-(dimethylamino)ethylmethacrylate) (PDMAEMA) or poly (2-(diethylamino)ethylmethacrylate) (PDEAEMA) to form nanoparticles of size 56.7 nm and with 2–3 nm pore sizes by TEM.[37] Transfection studies done in B16F10 cells showed that polycation-modified MSNs are able to simultaneously deliver chloroquine with DNA, and the co-delivery leads to a significantly increased transfection.

Published by Woodhead Publishing Limited, 2013

In order to reduce cytotoxicity and enhance transfection efficiency through receptor-mediated endocytosis, mannosylated PEI was coupled to the surface of MSN.[38] The size of the MSN/DNA complexes decreased from 130 to 60 nm, with an increase in weight ratio of 5 to 20. At a weight ratio of 20 the MSNs were spherical in shape with little aggregation and showed unimodal size distribution. Further, cell viabilities associated with treatment with MSN/DNA complexes were found to be higher than 80% up to a weight ratio of 20 in HeLa and Raw 264.7 macrophage cells. Also, MSN/DNA complexes showed enhanced transfection efficiency through receptor-mediated endocytosis via mannose receptors.[38]

16.6 Conclusions

Gold and mesoporous silica nanoparticles have emerged as promising vectors for gene delivery that provide a useful complement to more traditional delivery vehicles. The combination of low inherent toxicity, high surface area and tunable stability provides them with unique attributes that should enable new delivery strategies. However, there is a need to engineer nanoparticles with surface-optimized properties, such as bioavailability and non-immunogenicity, before advancement towards clinical applications.

16.7 References

1. Connor, E.E., Mwamuka, J., Gole, A., Murphy, C.J. and Wyatt, M.D. (2005) Gold nanoparticles are taken up by human cells but do not cause acute cytotoxicity. *Small*, **1**, 325–27.

Published by Woodhead Publishing Limited, 2013

2. Daniel, M.C. and Astruc, D. (2004) Gold nanoparticles: assembly, supramolecular chemistry, quantum-size-related properties, and applications toward biology, catalysis, and nanotechnology. *Chem Rev*, **104**, 293–346.
3. Love, J.C., Estroff, L.A., Kriebel, J.K., Nuzzo, R.G. and Whitesides, G.M. (2005) Self-assembled monolayers of thiolates on metals as a form of nanotechnology. *Chem Rev*, **105**, 1103–69.
4. Hostetler, M.J., Wingate, J.E., Zhong, C.-J., Harris, J.E., Vachet, R.W., et al. (1998) Alkanethiolate gold cluster molecules with core diameters from 1.5 to 5.2 nm: core and monolayer properties as a function of core size. *Langmuir*, **14**, 17–30.
5. Ghosh, P., Han, G., De, M., Kim, C.K. and Rotello, V.M. (2008) Gold nanoparticles in delivery applications. *Adv Drug Deliv Rev*, **60**, 1307–15.
6. Kim, C.K., Ghosh, P. and Rotello, V.M. (2009) Multimodal drug delivery using gold nanoparticles. *Nanoscale*, **1**, 61–7.
7. Brust, M., Walker, M., Bethell, D., Schiffrin, D.J. and Whyman, R. (1994) Synthesis of thiol-derivatised gold nanoparticles in a two-phase liquid-liquid system. *J. Chem. Soc., Chem. Commun.*, 801–2.
8. Schmid, G. (1992) Large clusters and colloids. Metals in the embryonic state. *Chem Rev*, **92**, 1709–27.
9. Templeton, A.C., Wuelfing, W.P. and Murray, R.W. (2000) Monolayer-protected cluster molecules. *Acc Chem Res*, **33**, 27–36.
10. Grabar, K.C., Freeman, R.G., Hommer, M.B. and Natan, M.J. (1995) Preparation and characterization of Au colloid monolayers. *Anal Chem*, **67**, 735–43.
11. Frens, G. (1973) Controlled nucleation for regulation of particle-size in monodisperse gold suspensions. *Nature Phys. Sci.*, **241**, 20–2.

Published by Woodhead Publishing Limited, 2013

12. Turkevich, J.S., Stevenson, P.C. and Hillier, J. (1951) A study of the nucleation and growth processes in the synthesis of colloidal gold, discuss. *Faraday Soc.*, **11** 55–7.
13. McIntosh, C.M., Esposito, E.A., Boal, A.K., Simard, J.M., Martin, C.T., et al. (2001) Inhibition of DNA transcription using cationic mixed monolayer protected gold clusters. *J Am Chem Soc*, **123**, 7626–9.
14. Sandhu, K.K., McIntosh, C.M., Simard, J.M., Smith, S.W. and Rotello, V.M. (2002) Gold nanoparticle-mediated transfection of mammalian cells. *Bioconjug Chem*, **13**, 3–6.
15. Han, G., Chari, N.S., Verma, A., Hong, R., Martin, C.T., et al. (2005) Controlled recovery of the transcription of nanoparticle-bound DNA by intracellular concentrations of glutathione. *Bioconjug Chem*, **16**, 1356–9.
16. Han, G., Martin, C.T. and Rotello, V.M. (2006) Stability of gold nanoparticle-bound DNA toward biological, physical, and chemical agents. *Chem Biol Drug Des*, **67**, 78–82.
17. Noh, S.M., Kim, W.-K., Kim, S.J., Kim, J.M., Baek, K.-H., et al. (2007) Enhanced cellular delivery and transfection efficiency of plasmid DNA using positively charged biocompatible colloidal gold nanoparticles. *Biochim Biophys Acta*, **1770**, 747–52.
18. Thomas, M. and Klibanov, A.M. (2003) Conjugation to gold nanoparticles enhances polyethylenimine's transfer of plasmid DNA into mammalian cells. *Proc Natl Acad Sci U S A*, **100**, 9138–43.
19. Wang, H., Chen, Y., Li, X.Y. and Liu, Y. (2007) Synthesis of oligo(ethylenediamino)-beta-cyclodextrin modified gold nanoparticle as a DNA concentrator. *Mol Pharm*, **4**, 189–98.

20. Uchimura, E., Yamada, S., Uebersax, L., Fujita, S., Miyake, M., et al. (2007) Method for reverse transfection using gold colloid as a nano-scaffold. *J Biosci Bioeng*, **103**, 101–3.
21. Hu, C., Peng, Q., Chen, F., Zhong, Z. and Zhuo, R. (2010) Low molecular weight polyethylenimine conjugated gold nanoparticles as efficient gene vectors. *Bioconjug Chem*, **21**, 836–43.
22. Cebrián, V., Martín-Saavedra, F., Yagüe, C., Arruebo, M., Santamaría, J., et al. (2011) Size-dependent transfection efficiency of PEI-coated gold nanoparticles. *Acta Biomater*, **7**, 3645–55.
23. Zhou, X., Zhang, X., Yu, X., Zha, X., Fu, Q., et al. (2008) The effect of conjugation to gold nanoparticles on the ability of low molecular weight chitosan to transfer DNA vaccine. *Biomaterials*, **29**, 111–17.
24. Han, G., You, C.-C., Kim, B.-J., Turingan, R.S., Forbes, N.S., et al. (2006) Light-regulated release of DNA and its delivery to nuclei by means of photolabile gold nanoparticles. *Angew Chem Int Ed*, **45**, 3165–9.
25. Rosi, N.L., Giljohann, D.A., Thaxton, C.S., Lytton-Jean, A.K.R., Han, M.S., et al. (2006) Oligonucleotide-modified gold nanoparticles for intracellular gene regulation. *Science*, **312**, 1027–30.
26. Giljohann, D.A., Seferos, D.S., Patel, P.C., Millstone, J.E., Rosi, N.L., et al. (2007) Oligonucleotide loading determines cellular uptake of DNA-modified gold nanoparticles. *Nano Lett*, **7**, 3818–21.
27. Radu, D.R., Lai, C.-Y., Jeftinija, K., Rowe, E.W., Jeftinija, S., et al. (2004) A polyamidoamine dendrimer-capped mesoporous silica nanosphere-based gene transfection reagent. *J Am Chem Soc*, **126**, 13216–17.
28. Suwalski, A., Dabboue, H., Delalande, A., Bensamoun, S.F., Canon, F., et al. (2010) Accelerated Achilles tendon

healing by PDGF gene delivery with mesoporous silica nanoparticles. *Biomaterials*, **31**, 5237–45.

29. Kim, M.-H., Na, H.-K., Kim, Y.-K., Ryoo, S.-R., Cho, H.S., et al. (2011) Facile synthesis of monodispersed mesoporous silica nanoparticles with ultralarge pores and their application in gene delivery. *ACS Nano*, **5**, 3568–76.
30. Gao, F., Botella, P., Corma, A., Blesa, J. and Dong, L. (2009) Monodispersed mesoporous silica nanoparticles with very large pores for enhanced adsorption and release of DNA. *J Phys Chem B*, **113**, 1796–804.
31. Kim, T.H., Kim, M., Eltohamy, M., Yun, Y.R., Jang, J.H., et al. (2012) Efficacy of mesoporous silica nanoparticles in delivering BMP-2 plasmid DNA for in vitro osteogenic stimulation of mesenchymal stem cells. *J Biomed Mater Res A*. doi: 10/1002/jbm.a.34466.
32. Yu, T., Greish, K., McGill, L.D., Ray, A. and Ghandehari, H. (2012) Influence of geometry, porosity, and surface characteristics of silica nanoparticles on acute toxicity: their vasculature effect and tolerance threshold. *ACS Nano*, **6**, 2289–301.
33. Zhu, S., Lu, H., Xiang, J., Tang, K., Zhang, B., et al. (2002) Novel nonviral nanoparticle gene vector: poly-l-lysine–silica nanoparticles. *Chin. Sci. Bull.*, **47**, 654–8.
34. Zhu, S.G., Xiang, J.J., Li, X.L., Shen, S.R., Lu, H.B., et al. (2004) Poly(L-lysine)-modified silica nanoparticles for the delivery of antisense oligonucleotides. *Biotechnol Appl Biochem*, **39**, 179–87.
35. Lunn, J.D. and Shantz, D.F. (2009) Peptide brush—ordered mesoporous silica nanocomposite materials. *Chem Mater*, **21**, 3638–48.
36. Hartono, S.B., Gu, W., Kleitz, F., Liu, J., He, L., et al. (2012) Poly-l-lysine functionalized large pore cubic

Published by Woodhead Publishing Limited, 2013

mesostructured silica nanoparticles as biocompatible carriers for gene delivery. *ACS Nano*, **6**, 2104–17.

37. Bhattarai, S.R., Muthuswamy, E., Wani, A., Brichacek, M., Castaneda, A.L., et al. (2010) Enhanced gene and siRNA delivery by polycation-modified mesoporous silica nanoparticles loaded with chloroquine. *Pharm Res*, **27**, 2556–68.
38. Park, I.Y., Kim, I.Y., Yoo, M.K., Choi, Y.J., Cho, M.-H., et al. (2008) Mannosylated polyethylenimine coupled mesoporous silica nanoparticles for receptor-mediated gene delivery. *Int J Pharm*, **359**, 280–7.

Published by Woodhead Publishing Limited, 2013

Index

Published by Woodhead Publishing Limited, 2013

Published by Woodhead Publishing Limited, 2013

Published by Woodhead Publishing Limited, 2013

CPSIA information can be obtained at www.ICGtesting.com
Printed in the USA
LVOW02*1425231013

358273LV00005B/11/P

9 781907 568404